POLYMER MATERIALS

Polymer Materials

an introduction for technologists and scientists

Second Edition

Christopher Hall

A HALSTED PRESS BOOK

JOHN WILEY & SONS
New York

First published 1989, Reprinted 1990 by

MACMILLAN EDUCATION LTD
London and Basingstoke

Published in the U.S.A. by
Halsted Press, a division of
John Wiley & Sons, Inc., New York

Printed in Hong Kong

ISBN 0–470–21092–3

Library of Congress Cataloging-in-Publication Data
Hall, Christopher, M. A.
    Polymer materials.
    "A Halsted Press book."
    Includes bibliographies and index.
    1. Polymers and polymerization.   I. Title.
TA455.P58H35   620.1'92   88-5520
ISBN 0–470–21092–3

Contents

Preface

Some systematic study of materials science now forms a part of many advanced courses in engineering and technology. There is no shortage of introductory textbooks of engineering materials science; yet most show a bias towards metals and ceramics and the treatment of polymers is often less thorough. On the other hand introductory books on polymer science have a preoccupation with polymerisation chemistry which makes them unsuitable for many students.

In this book I have attempted to provide a broad survey of the materials science of polymers for technologists, engineers and scientists, approximately at the level of general one-volume university texts on engineering materials. It is intended to serve both as a self-contained elementary treatment and also as a guide to the extensive specialist literature of polymer materials. I have sought to write a balanced scientific account with a somewhat technological flavour.

Nomenclature for polymers is carefully discussed in an appendix, following BSI and ASTM recommendations. The text makes extensive use of standard abbreviations such as PE, PA, PTFE and so on, but customary names such as acrylic and nylon have not been entirely excised.

In writing a book of broad compass, an author depends heavily on the established research and technical literature and on the indulgence of specialists. A number of such experts have kindly read parts of the manuscript in draft. The book is undoubtedly better for their comments, for which I am most grateful. I am particularly indebted to my colleagues Professor S. F. Bush and Dr W. D. Hoff for their careful reading of the text as it neared completion. I thank Mrs Jennifer Wilding for her help in preparing the typescript.

Manchester, 1980   C. HALL

Prefatory note to second edition

The aim of the second edition is the same as that of the first: to provide a broad scientific introduction to the material properties of polymers. In the seven years since the first edition appeared the specialist literature on polymer materials has grown hugely, increasing rather than reducing the need for introductory texts for students and non-specialists. At the same time, in science and engineering education, the integration of polymers into the main stream of materials teaching is gathering pace. I hope that this new edition will contribute to that process.

Cambridge, 1988   Christopher Hall

Acknowledgements

The author and publishers wish to thank the following who have kindly given permission for the use of copyright material:

Dow Chemical USA for figure 6.1;
Dr V. F. Holland and Monsanto Inc. for figure 2.5;
ICI Ltd for figure 4.5;
Professor A. Keller and Chapman & Hall Ltd for figure 2.6(d);
Dr A. Peterlin and the American Society for Metals for figures 2.7(c) and 3.3;
Dr A. Peterlin and Marcel Dekker Inc. for figure 2.7(b);
Dr A. Peterlin and Plenum Publishing Corporation for figure 2.7(a);
Mr V. Poirier and Thermo Electron Corporation for figure 5.9;
Mr. P. I. Vincent and the Plastics and Rubber Institute for table 3.2.

1
Polymers: Molecular Structure

A polymer is a very large molecule comprising hundreds or thousands of atoms, formed by successive linking of one or two, occasionally more, types of small molecules into chain or network structures. The concept of the polymer is one of the great ideas of twentieth century chemistry. It emerged in the 1920s amid prolonged controversy and its acceptance is closely associated with the name of Hermann Staudinger who received the Nobel Prize in 1953. The influence of the polymer (or *macromolecule*) concept spread rapidly into many areas of the natural sciences and technology. Within the life sciences it fostered the emergence of molecular biology through the study of natural macromolecular substances such as proteins, nucleic acids and polysaccharides. In engineering, a series of successes in commercial polymer synthesis established a new sector of the international chemical industry, devoted to producing and applying polymeric materials, notably plastics and rubbers, coatings and adhesives. This book is concerned with the materials science and engineering properties of such synthetic polymers.

1.1 The Polymer Materials Industry

Of the polymer materials in engineering use the plastics form the largest group by production volume. It is common to subdivide plastics into *thermoplastics* and *thermosets* (or thermosetting resins). Thermoplastics comprise the four most important commodity materials — polyethylene, polypropylene, polystyrene and poly(vinyl chloride) — together with a number of more specialised engineering polymers. The term 'thermoplastic' indicates that these materials melt on heating and may be processed by a variety of moulding and extrusion techniques. Important thermosets include alkyds, amino and phenolic resins, epoxies, unsaturated polyesters and polyurethanes, substances which cannot be melted and remelted but which *set* irreversibly. The distinction is important in that

1

production, processing and fabrication techniques for thermoplastics and thermosets differ.

Table 1.1 lists a number of plastics materials (a full discussion of polymer nomenclature appears in the appendix). The annual production of these materials in the United Kingdom and in the United States in the years 1985/1986 is shown in figure 1.1(a). The production of plastics, especially thermoplastics, demands a high level of chemical technology and is confined largely to a small number of major companies. A recent survey of the UK plastics industry showed that 90 per cent of thermoplastics production capacity is held by six companies. Table 1.2 charts the emergence of the major plastics materials.

Rubbers form another group of polymeric engineering materials. They are distinguished from plastics largely for reasons of industrial history. A rubber industry (using natural rubber latex as its raw material) was well established by

TABLE 1.1
Major polymer materials

Plastics – Thermoplastics	polyethylene	PE
	polypropylene	PP
	polystyrene	PS
	poly(vinyl chloride)	PVC
	polyacetal	POM
	acrylic	PMMA
	polyamide (nylon)	PA
	polycarbonate	PC
	polytetrafluorethylene	PTFE
Plastics – Thermosets	epoxy	EP
	melamine–formaldehyde	MF
	urea–formaldehyde	UF
	unsaturated polyester	UP
	phenolic	PF
	alkyd	
	polyurethane	PUR
Elastomers	natural rubber	NR
	styrene–butadiene rubber	SBR
	polybutadiene	BR
	butyl rubber	IIR
	polychloroprene	CR
	synthetic polyisoprene	IR
	nitrile	NBR
	silicone rubber	

TABLE 1.2
Emergence of some major plastics and rubbers

	Beginning of commercial production	
	year	country

Thermoplastics

Cellulose nitrate CN	1870	USA
Cellulose acetate CA	1905	Germany
Polystyrene PS	1930	Germany
Poly(methyl methacrylate) PMMA	1934	UK
Poly(vinyl chloride) PVC	1933	Germany/USA
Low density polyethylene LDPE	1939	UK
Polyamide PA	1939	USA
Polytetrafluorethylene PTFE	1950	USA
Acrylonitrile–butadiene–styrene ABS	1952	USA
Poly(ethylene terephthalate) PETP	1953	USA
High density polyethylene HDPE	1955	W. Germany
Polypropylene PP	1957	Italy
Polycarbonate PC	1959	W. Germany/USA
Polyoxymethylene POM	1960	USA
Polysulphone	1965	USA
Polymethylpentene	1965	UK
Linear low density polyethylene LLDPE	1977	USA

Thermosets

Phenol-formaldehyde PF	1909	USA
Urea-formaldehyde UF	1926	UK
Melamine-formaldehyde MF	1938	Germany
Polyurethane PUR	1943	Germany
Silicone SI	1943	USA
Polyester UP	1946	USA
Epoxy EP	1947	USA

Rubbers

Natural rubber NR (vulcanised)	1839	UK/USA
Styrene–butadiene rubber SBR	1937	Germany
Acrylonitrile-butadiene rubber NBR	1937	Germany
Polychloroprene CR	1932	USA
Polybutadiene BR	1932	USSR
Synthetic polyisoprene IR	1959	USA
Butyl rubber IIR	1940	USA
Ethylene-propylene rubbers EPM/EPDM	1963	USA/Italy

1900, some decades before the modern plastics industry, and before it was known that rubbers are polymeric substances. Today synthetic rubbers (*elastomers*) are widely used alongside natural rubber, figure 1.1(b), and a sharp distinction between plastics and rubbers is hard to sustain. Both are simply types of polymeric materials.

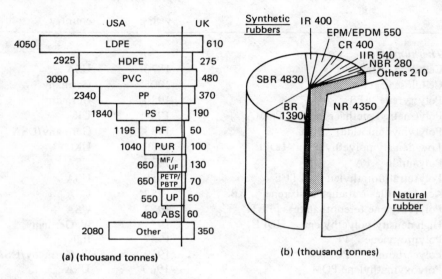

Figure 1.1 (a) United Kingdom and United States production of plastics materials compared, 1985–86 data; US production *per capita* is twice that of the UK. (b) Projected world consumption of rubber (Eastern Europe and parts of Asia excluded) for 1989 (International Institute of Synthetic Rubber Producers)

Similarly fibres, coatings and adhesives are polymeric materials designed to serve different ends and produced in different physical forms. Each is the concern of a particular industry, with a specialised technology. However the basic materials often have much in common. For example the nylons (or polyamides), important engineering thermoplastics, are found both in textiles and in coatings; epoxies are used both in paints and in adhesives and composites.

The structure of the polymer materials industry is summarised in figure 1.2. The polymers (with the exception of a few types produced by modification of vegetable substances such as cellulose and natural rubber) are produced from petroleum or natural gas raw materials. In the United Kingdom and Europe the key petrochemicals for polymer synthesis (ethylene, propylene, styrene, vinyl chloride monomer and others) are produced largely from naphtha, one of the distillation fractions of crude oil. In the United States natural gas provides the starting point. In both cases the polymer industry competes for its chemical feedstocks with other users of petroleum resources. Once synthesised, the polymer materials are passed to major consuming industries such as textiles or paints, or to a highly diverse processing sector, producing commodities for markets such as the building, packaging, agriculture, automobile, furniture, electrical and general engineering industries.

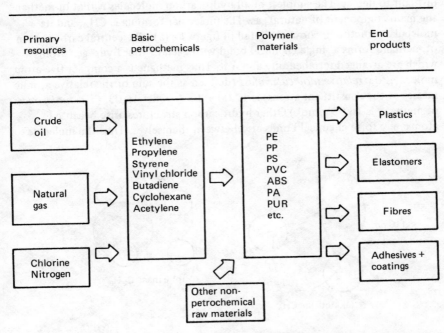

Figure 1.2   Production of polymer-based products from raw materials

Worldwide, the polymer materials industry continues to maintain the very high growth rate (perhaps 7 per cent per year overall) which has been evident for several decades. In the 1970s and 1980s, the industry has introduced many new *engineering polymers* and steadily improved the quality of its commodity products. In many applications, polymers are displacing other materials: for example, in the huge industrial and public utility piping market, where steel, cast iron, copper and fired clay are being displaced by polyethylene, polypropylene and poly(vinyl chloride) for water and gas distribution.

1.2 Hydrocarbons and Hydrocarbon Polymers

Hydrocarbons are a class of substances containing only the chemical elements carbon and hydrogen (C and H) in combination. Petroleum and natural gas are complex mixtures of hydrocarbons formed on Earth at remote times. A number of polymers (including some of the most important, PE, PP, PS and natural rubber) are also hydrocarbons: *hydrocarbon polymers*.

The difference between the individual hydrocarbon gases, liquids and solids (waxes and polymers) lies simply in *molecular structure* (figure 1.3). Fortunately since these structures involve only C and H atoms they are not difficult to depict. The simplest of all hydrocarbon molecules is that of methane, the main component of natural gas. The molecular formula is CH_4, and its molecular structure is shown in detail in figure 1.3(a). The central carbon atom of methane forms a single chemical bond with each of four hydrogen atoms, which are arranged tetrahedrally about it. Thus methane is a compact five-atom molecule. Its *relative molecular mass* (defined as the sum of the relative atomic masses of the constituent atoms) is $12.00 + 4 \times 1.008 = 16.03$. (The *molar mass* of methane is 16.03 g/mol.) Other hydrocarbon structures arise because C atoms may form chemical bonds also between themselves. Thus the molecule

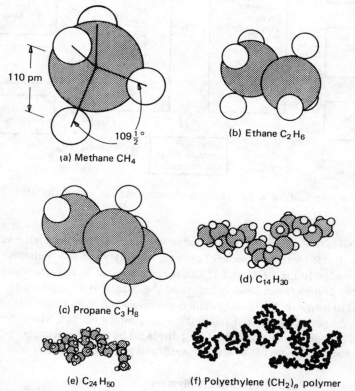

110 pm

$109\frac{1}{2}°$

(a) Methane CH_4

(b) Ethane C_2H_6

(c) Propane C_3H_8

(d) $C_{14}H_{30}$

(e) $C_{24}H_{50}$

(f) Polyethylene $(CH_2)_n$ polymer

Figure 1.3   (a) Molecular structure of methane CH_4 showing tetrahedral arrangement of C—H bonds, bond angle and bond length. (b) − (f) Molecular structures of other straight-chain hydrocarbons

of ethane (C_2H_6) comprises a pair of C atoms linked by a C—C single bond; each C atom is bonded in addition to three H atoms. Ethane is thus a C_2 hydrocarbon.

Higher hydrocarbons in the series are formed by extending the chain of C atoms, and paraffinic alkane hydrocarbons of this kind are found in natural petroleum oils up to a maximum carbon chain length of about C_{38}. There is a smooth change in the physical properties of the individual hydrocarbons as the chain length increases: thus the C_1–C_4 members of the series are gases at normal ambient temperature; C_5–C_{12} are volatile liquids, including important constituents of motor fuel (gasoline); C_{13}–C_{18} are higher boiling liquids (aviation fuel, kerosene); C_{19} and higher are heavy oils and waxes. Figure 1.4 shows how the important properties of melting point and density change with chain length n for the normal hydrocarbons C_nH_{2n+2}. The difference in properties between adjacent members of the series (C_n and C_{n+1}) becomes relatively smaller as C_n increases. This simply reflects the fact that as the chains get longer the addition of one further C atom represents a relatively smaller incremental change in molecular structure, and the effect of this change on properties becomes less marked. It becomes a very difficult task to separate the various components of the higher boiling fractions of petroleum from one another.

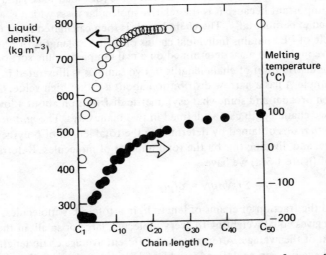

Figure 1.4 Liquid state densities and melting temperatures of normal straight-chain hydrocarbons C_nH_{2n+2} (densities at 20 °C or closest available temperature)

In 1933 a laboratory procedure was discovered which produced a synthetic hydrocarbon polymer of chain length far greater than C_{38}. The product was polyethylene (PE). PE is composed of hydrocarbon chains containing hundreds or thousands of carbon atoms. Thus we may write the molecular structure of PE as $H_3C(CH_2)_nCH_3$ or, omitting the terminal atoms, simply as $(CH_2)_n$. n is large, but we cannot assign it a unique value. The methods of synthesis of polymers normally produce a mixture of molecules of different chain lengths. As figure 1.4 suggests, such molecules differ only slightly in most physical properties and are not easily separated. We shall discuss the *distribution* of chain lengths and its consequences later.

PE resembles the paraffin waxes of $C_{30}-C_{40}$ in many respects: in appearance both are waxy, translucent, white solids, easily melted, of density about 900 kg/m^3 and both are electrical insulators. Both burn easily but are otherwise chemically inert. The most striking physical differences lie in mechanical properties, for whereas paraffin wax is mechanically weak, PE is a tough and useful engineering material.

1.3 Properties of the Polymer Chain

The *relative molecular mass* of polyethylene (*see* definition given in previous section) is

$$M = 14.02n + 2.02$$
$$\sim 14n \text{ when } n \text{ is large}$$

The relationship between relative molecular mass (or molar mass) and chain length is important because it is the relative molecular mass which is usually determined experimentally. The chain length is then calculated from this. Since any sample of PE contains individual chains of different lengths (n not constant) the relative molecular mass determined on a real sample is some sort of average value. The distribution of chain lengths in two samples is illustrated in figure 1.5(a). Sample A has a narrow distribution about a rather high value. Sample B has a much broader and somewhat asymmetric distribution about a lower value. The average chain length can be defined in two main ways. The *number-average chain length* $\bar{n}_N$ is obtained by determining the total length of polymer chain in the sample, and dividing this by the total number of molecules. Referring to curve A of figure 1.5(a) we have

$$\bar{n}_N = \Sigma f_i N n_i / N = \Sigma f_i n_i$$

where f_i is the fraction of chains of length n_i in a total of N molecules. This definition gives equal weighting to every molecule, large or small, in the calculation of the average. Alternatively a different average chain length, a *length-average* $\bar{n}_L$, can be defined in which chains contribute to the average in proportion to their length, the longer chains carrying greater (statistical) weight than the shorter ones.

$$\bar{n}_L = (\Sigma f_i N n_i \times n_i) / \Sigma f_i N n_i$$
$$= \Sigma f_i n_i^2 / \Sigma f_i n_i$$

Since the chain mass is directly proportional to the chain length, a *number-average relative molecular mass* $\bar{M}_N$ and a *weight-average relative molecular mass*

$\bar{M}_W$ are simply defined

$$\bar{M}_N = \bar{n}_N \times M_l$$

and

$$\bar{M}_W = \bar{n}_L \times M_l$$

where M_l is the relative molecular mass per unit length of the polymer chain. (In the SI system relative molecular mass and molar mass expressed in g/mol are numerically equal, so that $\bar{M}_N$ and $\bar{M}_W$ also define average molar masses.)

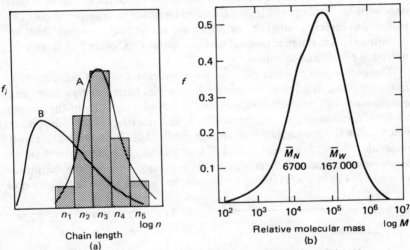

Figure 1.5 (a) Examples of narrow (A) and broad, asymmetric (B) polymer chain length distributions; curve A shown in histogram form (*see* text). (b) Experimental relative molecular mass distribution of an HDPE sample, showing $\bar{M}_N$ and $\bar{M}_W$; $f = n_i f_i / \Sigma(n_i f_i)$, obtained by size exclusion chromatography

The quantity $\bar{M}_W / \bar{M}_N$ (or $\bar{n}_L / \bar{n}_N$) equals 1 only for a polymer system in which all molecules have the same chain length and molecular mass. Such materials are called *monodisperse*. Otherwise $\bar{M}_W / \bar{M}_N > 1$ and this ratio is a measure of the broadness of molecular mass distribution — *see* figure 1.5(b). The relative molecular mass of a polymer has an important influence on flow properties in the molten state as well as on mechanical behaviour.

A polyethylene molecule containing, for example, 1000 C atoms has much the same length to thickness ratio as a piece of household string a couple of metres long. The maximum value of the end-to-end distance can easily be calculated from the C—C bond length which is accurately known to be 154 pm. Even in its fully extended conformation the molecule cannot attain a state in which all the C atoms are collinear, since the C—C bond angle is fixed at 109.5 degrees. However, a regular coplanar arrangement can be adopted, and is in fact found in the crystalline regions of solid PE. In such a conformation the end-to-end distance of a fully collinear 1000 C atom chain would be $999 \times 154 \times \sin(109.5°/2) = 0.13 \times 10^6$ pm = 0.13 μm.

Should we visualise the polymer chain molecule as a rigid or as a flexible entity? Since both the C—C bond angle and the bond length are fixed, chain flexibility can arise only from rotational motions about C—C bonds. Ethane is the simplest hydrocarbon which possesses a C—C bond, figure 1.3(b), and in the liquid and gaseous state the relative rotation of the CH_3 groups is known to be fairly free. The CH_3 rotors sense their orientation with respect to each other only weakly. Similarly the higher liquid and gaseous hydrocarbons may be regarded as loosely jointed chains, which are continually changing their conformations through rotational motions about C—C chain bonds, and the buffeting of thermal collisions.

This is true also of linear polymer chain molecules such as polyethylene. In the melt and in solution the molecule possesses kinetic energy distributed among many different kinds of motion (including translation of the molecule as a whole, rotation of parts of the molecule, and vibration of individual bonds), which cause the conformation to change continually. In these fluid states the molecule behaves as a loosely jointed chain and adopts a largely random arrangement in space, subject only to the bond angle restriction mentioned previously. Figure 1.6(a) shows a computer calculated projection based on a

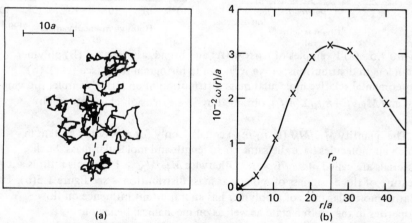

(a)   (b)

Figure 1.6 (a) Random conformation of a linear polymer chain of 500 bonds, with free rotation but fixed bond angle. The random chain is computed by successive addition of N vectors $\mathbf{a}_i$ of length a, randomly oriented in space subject to the condition that $\mathbf{a}_i \cdot \mathbf{a}_{i+1} = -a^2 \cos \alpha$, where $\alpha = 110$ degrees, the C—C—C bond angle. The figure shows the projection of the chain on to the plane of the paper. The parameter r, the end-to-end distance, is the length of the broken line. (b) r has an approximately Gaussian distribution $\omega(r)$ about a most probable value $r_p = (4Na^2/3)^{\frac{1}{2}}$

random walk model. Although very simple, this conveys a useful idea of the state of the random polymer chain in the melt and in solution.

In the solid state, the situation is less clear-cut. In some polymer solids the molecules adopt ordered crystalline arrangements; alternatively the solid polymer may be amorphous and lack long-range internal order. In either case, vibrational motion persists down to the lowest temperatures. Rotational motions also occur but are impeded by strong interactions between neighbouring molecules. Certain modes of molecular motion now require cooperation between molecular neighbours. The temperature is an important factor, for it determines how much kinetic energy the molecules possess. As the temperature falls various types of molecular motion are progressively frozen out. These effects underlie the mechanical behaviour of polymer materials and are discussed more fully in chapter 3.

1.4 Branched Chains

So far we have considered only linear polymer chains. We look now at the occurrence of branching during chain building. No such possibility arises in the case of the C_2 or C_3 hydrocarbons, but two C_4 alkane hydrocarbons can be envisaged

(I)   (II)

(I) is *n*-butane, $CH_3CH_2CH_2CH_3$ (the hydrogen atoms are omitted here to emphasise the chain backbone structure); (II) is isobutane, $CH(CH_3)_3$. (I) and (II) have different molecular structures, but they have the same number of C and H atoms in all, C_4H_{10}, and hence the same molar mass. They are butane *isomers*, distinct substances which differ in physical properties, such as melting points and boiling points: (I) $-0.5\,^\circ$C, (II) $-12\,^\circ$C. The physical differences arise simply from the different molecular *shapes*, illustrating the importance of chain branching.

Whereas there are only two C_4 alkane isomers, three C_5 isomers exist

Clearly as the number of C atoms in the chain increases, the number of branching options (and of isomers) rises rapidly. In fact the chains of polyethylene and other major thermoplastics are not highly branched. However a limited degree of chain branching is frequently found, and has important consequences. In PE produced by the high pressure process (*see* section 1.12) the molecules have (typically) short side branches about 4C long every 100C or so along the chain, and occasionally form long branches in addition. These branch points arise by molecular accidents during synthesis. The branch material is only a small proportion of the total material, but impedes crystalline packing of molecules. Moreover, points of weakness in the chain occur where branch and main chain join, and make the polymer vulnerable to degradation by ultraviolet light.

1.5 Stereoregularity

Polypropylene is another major commodity polymer closely related to polyethylene. It has the molecular structure

$$-\!\!\left(CH_2 - CH\right)\!\!_n$$
$$| $$
$$CH_3$$

and thus is formally derived from PE by the substitution of one of the H atoms on alternate C atoms of the chain by a CH_3 group. Polypropylene thus contains only C and H atoms and is a hydrocarbon polymer. Figure 1.7 shows the configuration of atoms in a short length of PE chain, and it is apparent that the PP structure can be derived from it in several ways. Structure (I) has all pendant CH_3 groups attached on the same 'side' of the chain; more precisely, all units have a spatially identical arrangement of atoms. Such a structure is called *isotactic.* Structure (II) on the other hand shows a random arrangement of pendant CH_3 groups in an *atactic* molecule. Structures (I) and (II) are distinct configurations, which cannot be interconverted by simple rotation about bonds. The *tacticity* or stereoregularity of polymer molecules is of profound importance for the properties of materials. It affects the way in which adjacent molecules can fit together in the dense packing of the solid and hence controls the strength of forces between molecules from which the mechanical properties of the material spring. The commercial form of PP is isotactic; atactic PP has no useful properties as a solid engineering material (*see also* section 6.1).

Tacticity arises in all chain polymers in which atoms of the backbone carry two dissimilar atoms or groups. Thus commercial PS is atactic; PMMA is largely *syndiotactic*, that is, the configuration at the C atom carrying the side groups shows a regular alternation along the chain.

1.6 Other Hydrocarbon Polymers

There are several other important polymers which like PE and PP are based on

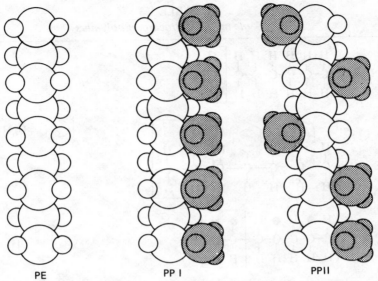

Figure 1.7   Relationship between the molecular structures of polyethylene (PE) and the isotactic (I) and atactic (II) forms of polypropylene (PP)

C—C linear chains and contain only C and H atoms. Table 1.3 shows that the structures can easily be understood by reference to PE.

1.7 Other Carbon Chain Polymers

Atoms of other elements may be incorporated into polymer structures (table 1.4). Substitution of H by chlorine Cl or fluorine F leads to several major polymers, PVC, PTFE, PVDC, PCTFE. The cyano group CN may similarly be incorporated, as in PAN. More complicated substituents composed of groupings of H, C and O atoms also occur, as in PMMA, PVAC, PVAL and others.

1.8 Heterochain Polymers

All the polymer chain structures listed in tables 1.3 and 1.4 are based on C chain backbones. The replacement of C atoms in the backbone itself by atoms of other elements produces *heterochain* polymers. A structurally simple heterochain polymer is polyoxymethylene (POM)

$$-CH_2-O-CH_2-O-CH_2-O-$$

which may be thought of as a PE chain in which O atoms replace alternate CH_2 groups. POM and closely related materials are important engineering plastics

TABLE 1.3
Some major hydrocarbon polymers

PE

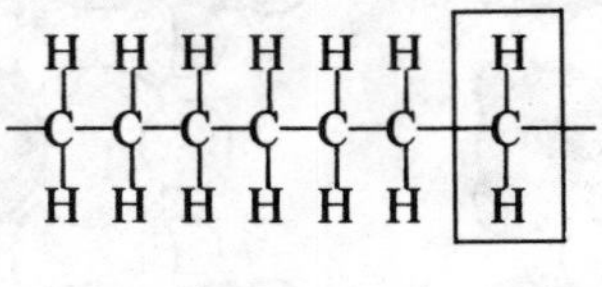

PS

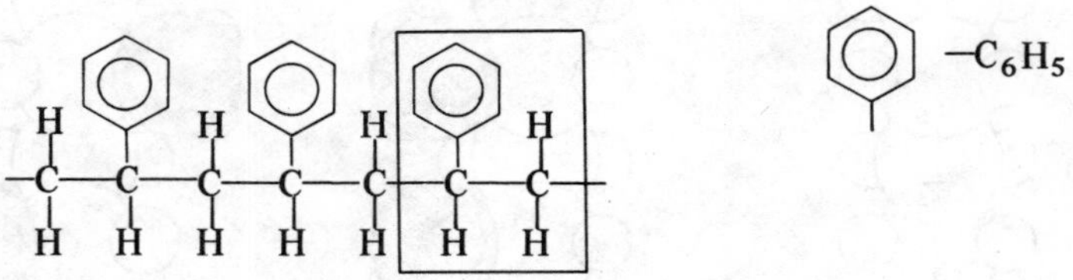

$-C_6H_5$

PP

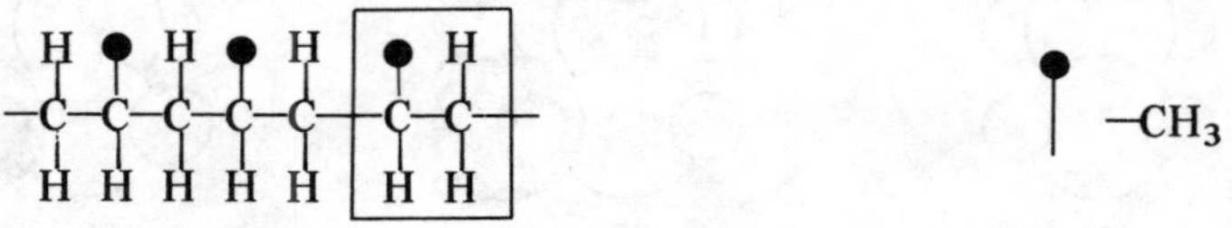

$-CH_3$

NR

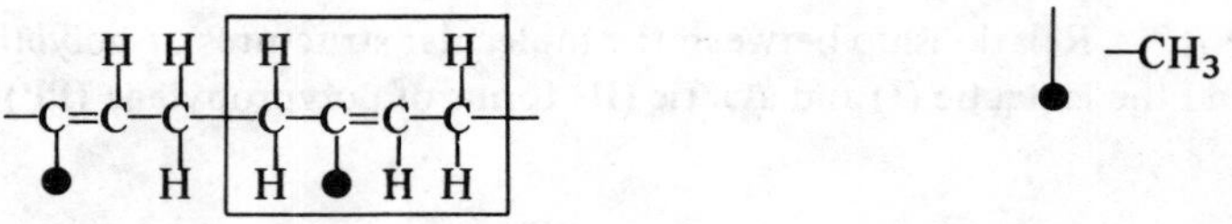

$-CH_3$

Polymethylpentene

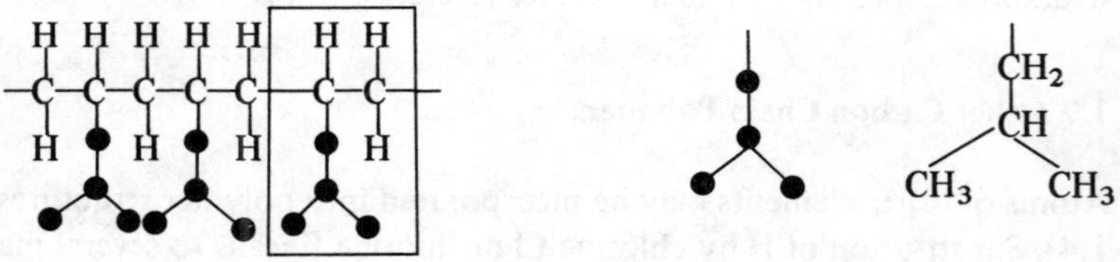

CH_2
CH
$CH_3 \quad CH_3$

BR

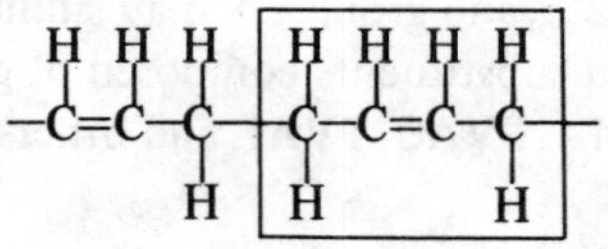

(commonly known as acetals). Other heterochain polymers are listed in table 1.5 and include the PA group, the PUR group and PETP. Incorporating heteroatoms in the chain generates great structural diversity.

A number of *inorganic* polymers exists in which the backbone contains no carbon atoms at all. Pre-eminent amongst these are the polysiloxanes, linear Si—O polymers which have important properties of inertness and heat resistance.

TABLE 1.4
Some important carbon chain polymers

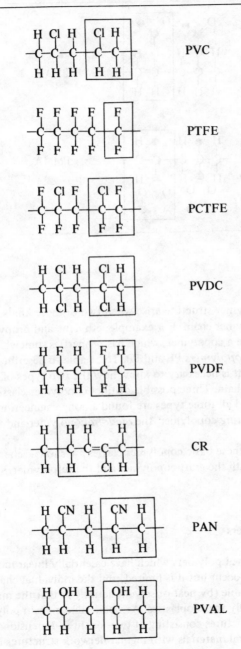

TABLE 1.4 continued

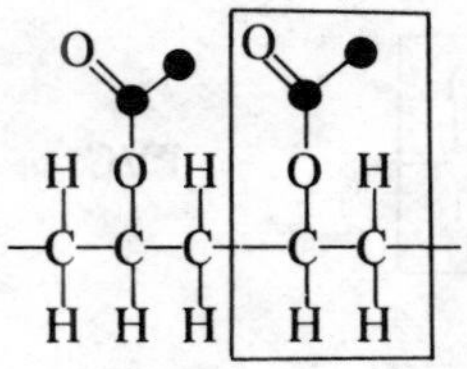

PVAC

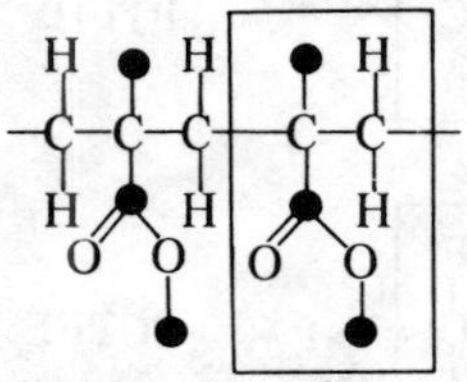

PMMA

1.9 Copolymers

Further types of polymer structure arise when two or more kinds of *mer* are mixed in a single polymer chain. For example, ethylene and propylene may be *copolymerised* to give a copolymer, which has properties somewhat different from the parent *homopolymers* PE and PP. To be able to describe the copolymer chain structure fully it is necessary to know how the two types of mer are arranged within the chain. Three possible distinct copolymer classes can be envisaged at once, and all three types are found among engineering polymer materials: the *alternating* copolymer, the *random* copolymer and the *block* copolymer.

The structures of these basic copolymer classes are shown schematically in table 1.6, together with the graft copolymer and the terpolymer.

1.10 Network Polymers

So far we have discussed polymers which have essentially linear molecular chains. Some branching may occur but it is limited, and the individual chain molecules are distinct and separable (by heat or solvent action) so that the materials we have listed are generally thermoplastics. As well as linear chain polymers there are materials with structures consisting of two- or three-dimensional networks of chemical bonds. Several materials with regular network structures lie on the

TABLE 1.5
Some important heterochain polymers

Type	Characteristic chain group	Example	

Type	Characteristic chain group	Example
Polyether	$-\overset{\|}{\underset{\|}{C}}-O-\overset{\|}{\underset{\|}{C}}-$	PEO $\left[-\overset{H\ H}{\underset{H\ H}{C-C}}-O-\right]$ $H-\overset{H}{\underset{H}{C}}-H$
		PPO
Polyamide	$-\overset{O}{\overset{\|\|}{C}}-\overset{H}{\underset{\|}{N}}-$	PA6 $\left[-\overset{H\ H\ H\ H\ H\ O}{\underset{H\ H\ H\ H\ H}{C-C-C-C-C-C}}-\overset{H}{N}-\right]$
Polyester	$-O-\overset{O}{\overset{\|\|}{C}}-$	PETP $\left[-O-\overset{H\ H}{\underset{H\ H}{C-C}}-O-\overset{O}{\overset{\|\|}{C}}-\bigcirc-\overset{O}{\overset{\|\|}{C}}-\right]$
		PC
Polyurethane	$-O-\overset{O}{\overset{\|\|}{C}}-\overset{H}{\underset{\|}{N}}-$	$\left[-O-\overset{O}{\overset{\|\|}{C}}-N-\bigcirc-N-\overset{O}{\overset{\|\|}{C}}-O-\sim\right]$ ~~~ a linear polymer chain
Polysulphide	$-C-S-S-$	$\left[-\overset{H\ H}{\underset{H\ H}{C-C}}-O-\overset{H}{\underset{H}{C}}-O-\overset{H}{\underset{H}{C}}-\overset{H}{\underset{H}{C}}-S-S-\right]$
Polysulphone	$-\overset{O}{\underset{O}{\overset{\|\|}{\underset{\|\|}{S}}}}-$	$\left[-O-\bigcirc-C-\bigcirc-O-\bigcirc-S-\bigcirc-\right]$
Silicone (siloxane)	$-Si-O-$	$\left[-Si-O-\right]$
Polyimide	(imide structure)	(polyimide structure)

17

TABLE 1.6
Classification of copolymers

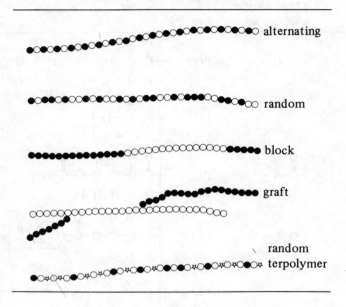

borderline between polymers and ceramics (figure 1.8). For example, regular two-dimensional networks occur in graphite and boron nitride; diamond has a regular three-dimensional lattice. Many silicate minerals are based on polymeric Si—O layer and lattice structures.

Most common network polymers are not regular in structure. Such materials are generally thermosets or elastomers, since the entire network is interconnected through primary chemical bonds and distinct, separable molecules do not exist. There are two principal types of network materials

(1) Those formed by the linking of linear chains by small molecules.
(2) Those formed by the direct reaction of small molecules (including short prepolymer chains) which give rise to chain-branching.

The first type includes many elastomers. For example, raw natural rubber is a hydrocarbon polymer, polyisoprene, which contains reactive C=C bonds. In vulcanising rubber, molecules of sulphur form crosslinks between C=C bonds on adjacent molecules. In a similar way, unsaturated polyesters UP can be crosslinked by molecules such as styrene to produce network polymers. The second type is represented by important thermosets such as PF, MF and EP (table 1.7). The structure of the polymerised material is highly complicated but figure 1.9(a) shows the type of network which is formed.

TABLE 1.7
Synthetic network polymers (thermosets)

Type	Starting materials	Network polymer

PF

phenol   formaldehyde

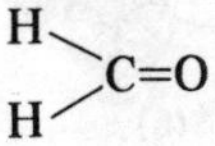

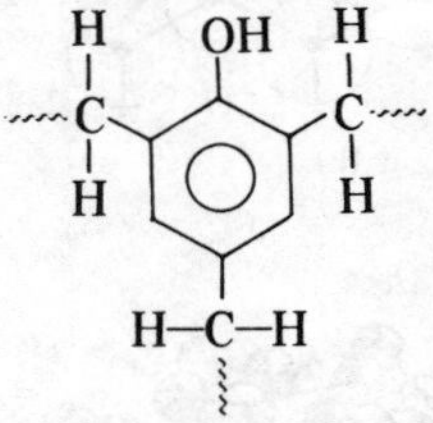

and similar units randomly connected by a variety of links

MF

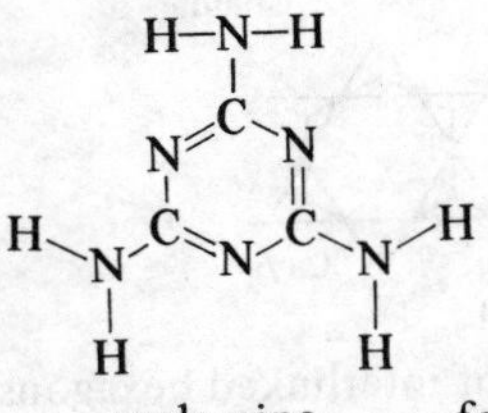

melamine   formaldehyde

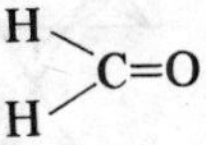

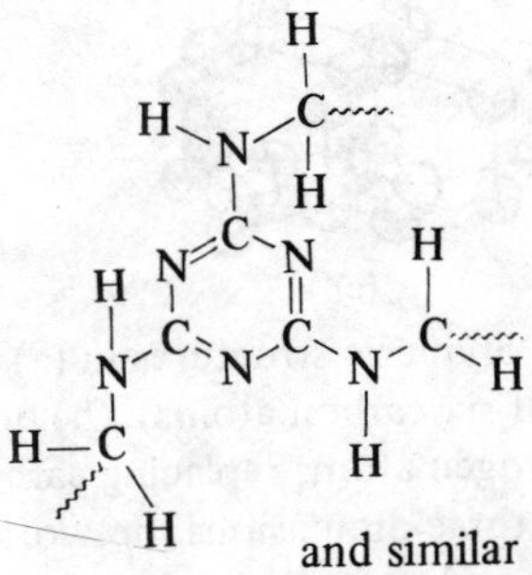

and similar units randomly connected by a variety of links

EP

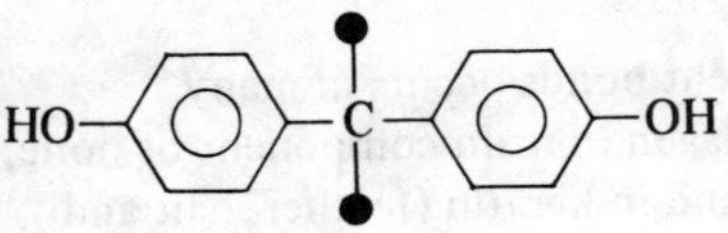

bisphenol A

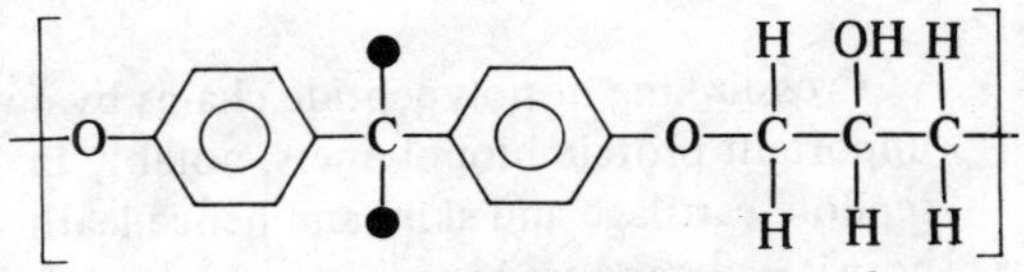

crosslinked through terminal epoxy groups

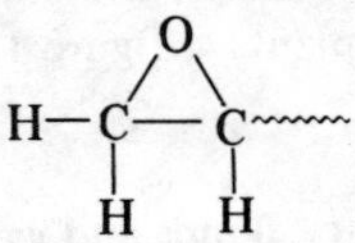

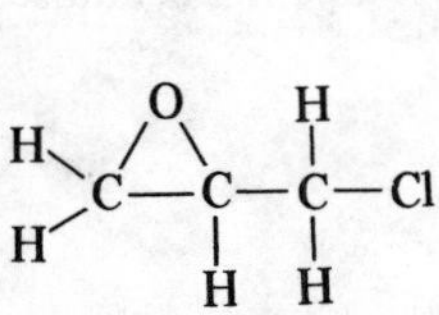

epichlorhydrin

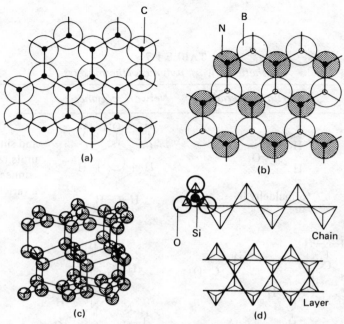

Figure 1.8   Polymeric structures in (a) graphite (sheets of interlinked hexagons, each formed of six carbon atoms); (b) boron nitride (graphite-like structure with boron and nitrogen atoms replacing carbon); (c) diamond (another form of carbon with a three-dimensional network of C—C bonds arranged tetrahedrally); and (d) chain and layer silicates based on the SiO_4 tetrahedron (including minerals such as asbestos and talc)

Crosslinking of polypeptide chains by covalent bonds occurs in many important protein biopolymers, notably in collagen (a main component of bone, tendon, cartilage and skin, and hence leather) and in keratin (feather, hair and horn). Few random branched polymers of the second type occur in biology, but one major example is lignin, the structural material which accompanies cellulose in woody plants – figure 1.9(b).

1.11 Primary Bonds and van der Waals' Forces

The polymer molecular structures we have described are more or less complicated arrangements of atoms held together by primary chemical bonds. Relatively few elements appear in the structures of the common polymers.

The principal chain building bonds are C—C, C—O and C—N; and C—H, C—F, C—Cl, C=O, O—H and N—H occur frequently, attached directly or indirectly to

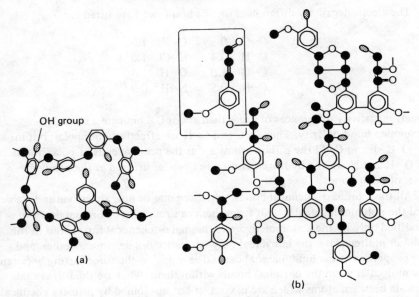

Figure 1.9 (a) Random network structure of a PF thermoset. (b) Proposed illustrative structure for lignin (Adler): random network polymer based on phenylpropane unit, top left

the chain backbone. All these chemical bonds are formed by rearranging valency electrons from the outer regions of the participating atoms into new spatial distributions which embrace both atoms. If the two atoms forming the bond are dissimilar (as in all those listed except C—C) the sharing of the valency electrons is unequal. The electron distribution is not symmetrical and the bond has a definite electrostatic polarity, which arises because one atom is able to draw electron charge more strongly to itself than the other. A measure of the electron-attracting power of an atom in a chemical bond is the *electronegativity* of the element, and the electronegativity difference of the two atoms of the bond is a good guide to bond polarity. The electronegativities of the main elements appearing in polymer materials are

C	2.5
H	2.1
N	3.0
O	3.5
F	4.0
Cl	3.5

The electronegativity differences for the bonds we have listed are

C—C	0	C—F	1.5
C—H	−0.4	C—Cl	1.0
C—O	1.0	O—H	1.4
C—N	0.5	N—H	0.9

Electronegativity differences of more than about 0.8 indicate a strongly developed bond polarity. Thus C—C and C—H are effectively nonpolar; but in C—O, C—F and C—Cl the carbon atoms are at the positive ends of polar bonds. In O—H and N—H, the hydrogen atoms are likewise situated at the positive ends of polar bonds.

All these primary chemical bonds, whether polar or not, are strong and stable, and the polymer molecules built from them can generally be broken down or modified only by the action of vigorous thermal or chemical forces. But in the solid or molten state, the individual polymer molecules lie close together, and the properties of the bulk material depend as much on the forces acting between the molecules as on the chemical bonds within them. What are the forces that operate between atoms which are in contact but not joined by primary chemical bonds? Collectively these secondary interactions are known as van der Waals' forces. Unlike primary chemical bonding, their operation does not involve sharing or transfer of electrons between atoms. The weakest of the interactions is a feeble force of attraction known as a dispersion force which exists between all atoms as the result of rapid fluctuations of the atomic electron distributions. The dispersion force between each pair of adjacent —CH_2— mers on two polyethylene chains lying side by side is about 200 times weaker than the force acting within each C—H bond.

Somewhat stronger than the dispersion force is the van der Waals' force acting between suitably oriented polar bonds. A particularly strong and important type of polar force is the *hydrogen bond*, found in several synthetic polymer materials and widely in biopolymers. The hydrogen bond is formed between a markedly electronegative atom and a hydrogen atom already bonded to another electronegative atom − figure 1.10(a). Thus O—H . . . O hydrogen bonds occur in PVAL and cellulose. N—H . . . O hydrogen bonds are formed between PA molecules − figure 1.10(b), and in proteins, including the fibrous proteins such as wool keratin and silk fibroin. (N—H. . .O hydrogen bonds also link the two antiparallel macromolecular chains forming the double helix of DNA, so that the properties of the hydrogen bond lie at the roots of molecular biology). The strength of a hydrogen bond depends on the identity of the electronegative atoms involved and on their exact relation to other atoms, but it is often as great as one-tenth of the strength of a typical primary chemical bond. Hydrogen bonds are frequently sufficiently strong to survive the forces arising from molecular thermal motion at normal temperatures. We return to the role of intermolecular forces in polymer materials in chapter 2.

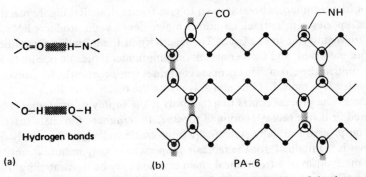

Figure 1.10   The hydrogen bond. (a) O—H. . .O and N—H. . .O hydrogen bonds.
(b) H-bond formation between extended PA- 6 chains; similar hydrogen bonds
occur in other polyamides

1.12 Polymer Synthesis: Chain Reaction Polymerisation

We conclude this chapter with a brief survey of the chemical reactions in which
polymer chains are formed from substances of low molar mass. The treatment is
brief because polymer synthesis is a somewhat specialised topic for the polymer
chemist and the chemical engineer. A fuller discussion would call for a greater
knowledge of the principles of chemical reactivity on the part of the reader than
has been assumed. Many excellent texts on polymerisation chemistry are
available, and several are listed at the end of the chapter.

In the first laboratory synthesis of polyethylene the hydrocarbon gas ethylene
(ethene) was heated to 170 °C at a pressure of 1300 bar. The white solid formed
in the reaction vessel was rapidly and correctly identified as a polymer of
ethylene. The ethylene had reacted according to the equation

$$n/2 \; C_2H_4 \longrightarrow \; (CH_2)_n$$

This is an example of one of the two major classes of polymerisation reaction,
the *addition polymerisation*, which may be written more generally

$$nM \longrightarrow M_n$$

In addition reactions the polymer is the sole product of the reaction. Addition
polymerisation almost invariably occurs by a *chain reaction* mechanism,
frequently involving free radicals. Such reactions, which are of great industrial
importance, proceed according to the following scheme

$$I \longrightarrow 2R\cdot \qquad \text{initiation}$$

$$R\cdot + M \longrightarrow RM\cdot \qquad \text{propagation}$$

$$RM\cdot + M \longrightarrow RMM\cdot \qquad \text{etc.}$$

$$RM\!\sim\!\! M\cdot + RM\!\sim\!\! M\cdot \; \rightarrow \; RM\!\sim\!\! MM\!\sim\!\! MR$$
$$\text{termination}$$

An initiator molecule breaks down to give free radicals $R\cdot$, highly reactive chemical species which attack monomer molecules to yield products $RM\cdot$, etc., which are themselves free radicals. The reactivity of the initiating free radical is therefore preserved, and further attack on a monomer molecule occurs to add a further unit to the chain. This process continues (*propagation*), the growing chain with a free radical at one end scavenging the reaction medium for monomers. The process comes to a halt only if the supply of monomer is exhausted or if free radicals combine together to *terminate* the chain reaction. The latter process is usually statistically improbable so long as the supply of monomer is maintained, for the radicals are present in only minute quantities.

The mechanism of a free radical chain reaction may be illustrated by the example of the polymerisation of vinyl chloride. A variety of initiators may be used, for example lauroyl peroxide

$$\text{Initiator I} \longrightarrow 2R\cdot$$

$$R-(CH_2-CHCl)_n-CH_2-\dot{C}HCl + CH_2{=}CHCl$$

$$\longrightarrow R-(CH_2-CHCl)_{n+1}-CH_2-\dot{C}HCl \qquad \text{etc.}$$

Propagation

$$R-(CH_2-CHCl)_n-CH_2-\dot{C}HCl$$
$$+ R-(CH_2-CHCl)_m-CH_2-\dot{C}HCl$$

$$R-(CH_2-CHCl)_n-CH_2-CHCl-CHCl-CH_2-(CHCl-CH_2)_m-R$$

Combination termination

$$R-(CH_2-CHCl)_n-CH{=}CHCl + R-(CH_2-CHCl)_m-CH_2-CH_2Cl$$

Disproportionation termination

Each initiator is responsible for the formation of one polymer chain. The rate of growth of a chain is roughly constant as it is largely determined by chemical factors which change little as propagation proceeds. Towards the end of the reaction the rate of polymerisation may be affected by depletion of monomer and by an increase in viscosity caused by the accumulation of the long polymer chains themselves.

Once initiated, polymer chains usually grow very rapidly until a chain termination occurs and growth abruptly ceases. The average final length of the chain depends on the relative probabilities of propagation and termination. As the bulk reaction proceeds, the number of polymer chains increases steadily, but the average length of the chains is roughly constant. These features of chain growth polymerisation reactions may be contrasted with the characteristics of the step reactions described below.

The high reactivity of the free radical species (both the initiators themselves and the active centre on the growing chain) means that there is often a tendency for side reactions to occur. These side reactions may influence the degree of polymerisation which occurs in a practical synthesis, as well as the molar mass distribution of the product.

Free radicals are not the only reactive chemical substances which can act as initiators in chain polymerisations. The active centre in the propagating polymer may be an ion rather than a free radical. Generally the character of the polymerisation is similar to that described above. One class of such reactions is especially important because the tacticity or stereoregularity of the polymer product is controlled. Stereospecific polymerisations can be achieved by using a wide variety of reactive initiators and catalysts, particularly the *Ziegler–Natta catalysts.*

In order to produce a stereoregular polymer such as isotactic polypropylene the orientation of the monomer molecule has to be firmly controlled at the point at which it is added to the growing chain. Ziegler–Natta catalysts achieve this by pinning the active end of the chain to the solid catalyst surface throughout the polymerisation. This restricts the direction of approach of the incoming monomer and orients it in relation to the chain.

Ziegler–Natta catalysts (of which very many variants exist, see figure 1.11) have made it possible to synthesise several stereoregular polymers of commercial importance, particularly PP and the rubbers polyisoprene (IR) and polybutadiene (BR). Ziegler catalysts are also used in low pressure processes to produce the linear form of PE known as high density polyethylene (HDPE). Karl Ziegler and Giulio Natta shared the 1963 Nobel Prize for chemistry for their work on stereoregular polymerisation. Other important low pressure catalytic routes to HDPE have subsequently been introduced by Phillips Petroleum and Standard Oil of Indiana.

1.13 Step Reaction Polymerisation

The second major class of polymerisation reaction, the condensation reaction occurring by a step reaction mechanism, may be illustrated by the formation of the polyamide from an amino acid

$$n\text{NH}_2(\text{CH}_2)_{10}\text{CO}_2\text{H} \longrightarrow \{\text{NH}(\text{CH}_2)_{10}\text{CO}\}_n + (n-1)\text{H}_2\text{O}$$

Condensation polymerisations have the following general form

$$n\text{MN} + n\text{PQ} \longrightarrow (\text{MP})_n + n\text{NQ}$$

MN and PQ may be the same, as in the example given, or may be different. Thus polyamides are often formed by the condensation of a diacid and a diamine,

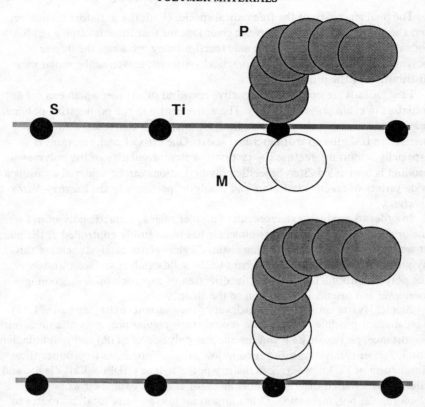

Figure 1.11 A typical high-activity Ziegler–Natta catalyst is made by reaction of titanium tetrachloride, magnesium chloride and triethylaluminium, forming a solid catalyst surface S. A free monomer molecule M bonds briefly to a titanium ion Ti exposed at the catalyst surface and to which the end of the growing polymer chain P is already attached. A spontaneous and rapid rearrangement of chemical bonds then occurs to insert the monomer into the chain. The polymerisation continues until a termination reaction eventually intervenes. It is the geometry of the active site on the catalyst surface which provides the remarkable steric control as the polymer chain grows. The configuration of the catalytic site, acting as a template, ensures that each successive reacting monomer is identically oriented in relation to the end of the polymer chain

for example

$$nH_2N(CH_2)_6NH_2 + nHO_2C(CH_2)_4CO_2H \longrightarrow$$
$$H[NH(CH_2)_6NHCO(CH_2)_4CO]_nOH + (2n-1)H_2O$$

This necessarily generates an alternating copolymer, although the product is not usually so regarded. The polymer itself is not the sole product of a condensation polymerisation, a major point of difference between addition and condensation reactions. In the examples above, one molecule of water (H_2O) is formed for every pair of reactant molecules which combine. The term 'condensation' is used generally in organic chemistry to refer to reactions in which molecules combine together through the elimination of a small molecule such as water.

There are several contrasts between step and chain polymerisation reactions. Step reaction monomers do not require activation by an initiator but are intrinsically reactive. Since initiation is not required, all monomers present at the start of the reaction have an equal chance of reacting. In the early stages of the reaction, monomers combine to form short chains. Since the rate of these step reactions is typically much slower than the rate of the propagation stage in chain reactions, long chains do not develop at all in the early stages. As the amount of free monomer in the system diminishes, the step reaction occurs increasingly between functional groups on the ends of chains. As the reaction progresses, there is an increase in the average chain length from initially low values to ultimately large values. To obtain very long chains it is necessary that the step reactions should proceed almost to completion. It is in the last stages of the reaction when links are being forged between already well-developed chains that the average chain length rises most rapidly.

The mechanisms and kinetics of polymerisation reactions were intensively investigated in the period from about 1928 until the 1950s, culminating in 1953 in the publication of Paul Flory's book, *The Principles of Polymer Chemistry*, one of the classic works of polymer science. Flory received the Nobel prize in 1974, the fourth and most recent Nobel award in the field of synthetic polymers. The standard theory of polymerisation provides a quantitative description of both step and chain reactions. In particular, it successfully predicts the way in which the rates of polymerisation reactions depend on the concentrations of reacting monomers and how the distributions of chain lengths alter during the course of these reactions. This understanding of course is essential to the design of all industrial polymer syntheses.

1.14 Thermoset Polymerisation

If one of the reacting substances in a step polymerisation possesses more than two functional groups, then the reaction will lead naturally to branched structures (table 1.8). As these intercombine, random three-dimensional networks are developed. Ultimately the network extends throughout the mass of polymerised material, which therefore in effect constitutes a single giant molecule. Such structures are by their very nature incapable of melting or of truly dissolving. Polymer materials of this type are called *thermosets*, since once produced they cannot be returned to a fluid condition by heating. Important

TABLE 1.8
Formation of linear and branched chains

Linear chain formation:

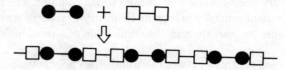

Branched chain (network polymer) formation:

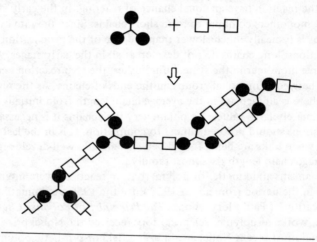

examples are the phenolic and amino resins. Similar thermoset network structures may be formed by crosslinking linear chains in a second post-polymerisation or 'curing' reaction, which may be free radical initiated. Tables 1.9 and 1.10 show how thermosets derived from unsaturated polyesters and epoxides are produced by two-stage reactions — curing following resin (prepolymer) formation. The irreversibility of set (and, frequently, stiffness and good resistance to heat) underlies the great technical value of thermosets, notably as adhesives and as the bonding component of fibre-reinforced and mineral-filled composite materials. A number of important surface-coating polymers (notably polyurethanes and alkyds) are thermosets.

1.15 Polymerisation Conditions

We have mentioned earlier that such important properties of the polymer product as chain length (and its distribution) and stereoregularity are determined by details of the synthetic method employed. Other variables which we have not

TABLE 1.9
Preparation and curing of unsaturated polyesters

Typical starting materials:

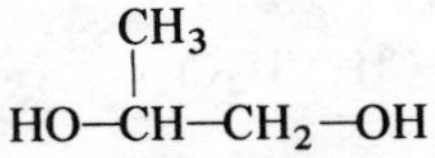

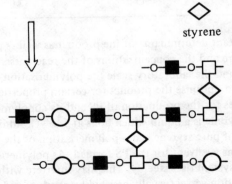

$$CH_3$$
$$HO{-}CH{-}CH_2{-}OH$$

propylene glycol 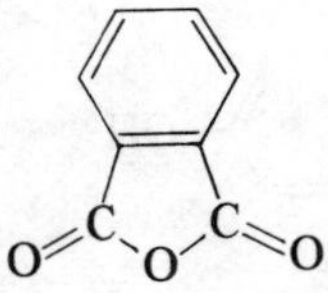 · · · · · · · · · phthalic anhydride 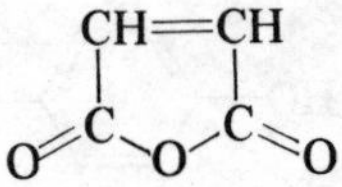 · · · · · · maleic anhydride

We represent these molecules by the symbols

HO—■—OH

Polymerisation produces a linear unsaturated polyester with a structure like this:

—O—■—O—◯—O—■—O—□—O—■—O—□—

The addition of a vinyl monomer such as styrene makes a liquid resin which can be crosslinked by a free-radical initiator (*curing agent*)

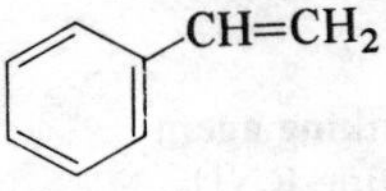

$$CH{=}CH_2$$

◇

styrene

Schematic crosslinked polyester structure

TABLE 1.10

Preparation and crosslinking of an epoxy resin

Starting materials:

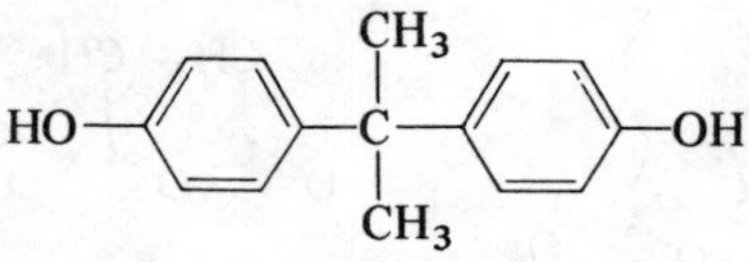

phenol (bisphenol A) 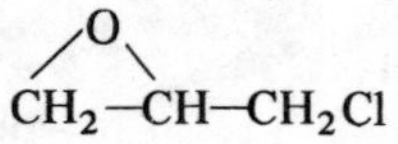   epoxide (epichlorhydrin)

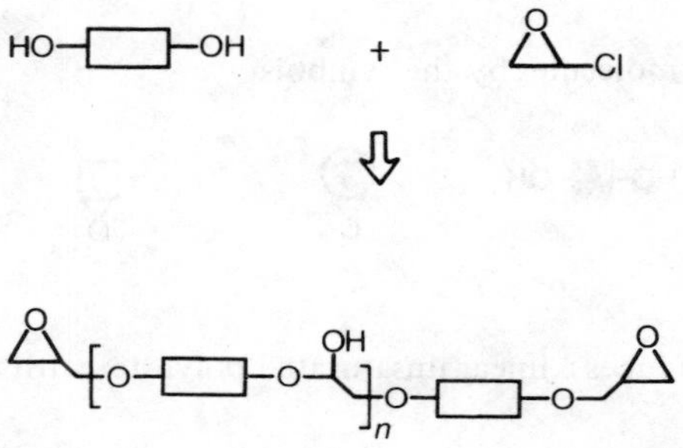

$n = 1-12$

Crosslinking occurs through the terminal epoxy groups:

Crosslinking agent, e.g. amine RNH_2

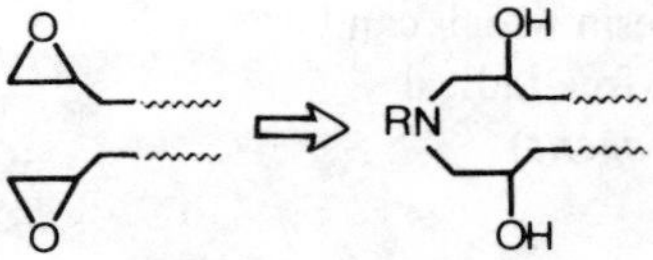

discussed are important determinants of the product as well — parameters such as temperature, pressure and the concentrations of the reactants and catalysts. On both the industrial and the laboratory scale the polymerisation conditions may be chosen in order to optimise the product for certain properties. Consider the commercial processes for the production of PS and its copolymers.

The monomer styrene is a liquid which boils at 146 °C. Addition of an initiator to a batch of pure styrene starts polymerisation of the entire mass of liquid. Styrene acts as a solvent for the PS formed. As polymerisation proceeds, the viscosity of the system rises steeply. This may interfere with heat loss to the walls of the reaction vessel (usually assisted by stirring) and cause local hot spots to develop. In these regions the reaction occurs in an uncontrolled fashion and this may give rise to undesirable changes in the quality of the product. This method of producing PS is an example of a *bulk polymerisation* operated batchwise. It has some commercial application in the production of PS in block form in special reactors designed for efficient heat removal. Since the monomer is used in pure form the product is of high purity. Bulk polymerisation of styrene is more commonly carried out in flow reactors which receive unpolymerised styrene feed at about 100 °C. Polymerisation occurs as the styrene passes through the reactor, and to control the rising viscosity the temperature is increased steadily to about 200 °C at the PS outlet. (Polymers are not invariably soluble in their monomers; for example, the bulk polymerisation of PVC leads to precipitation of polymer as the reaction progresses.)

A solvent such as ethylbenzene may be added to the styrene feed, converting the bulk process to a *solution polymerisation*. Commercial processes for the solution polymerisation of PS or copolymers such as SAN use only small amounts of solvent. Solution processes are used in producing many polymers, since good control of viscosity and heat transfer can be obtained. The polymer product may be either soluble or insoluble in the selected solvent. It is necessary to remove the solvent to obtain the pure polymer but this is often difficult to do completely. If the polymer is insoluble in the solvent, it is precipitated as it is formed. Such methods lend themselves to continuous production processes, for the solid polymer can be continuously removed from the reaction medium and the monomer and initiator replenished. However, the presence of large amounts of solvent increases the risk of side reactions, such as chain transfer, which prematurely terminate chain growth — a factor which influences the choice of solvent.

Two principal alternatives to the single-phase polymerisations exist. Styrene is not appreciably soluble in water. A mechanically agitated mixture of water and styrene (2 + 1) forms a dispersion of styrene droplets in water, which may be stabilised by a small amount of a colloidal additive. In the *suspension* process, initiator is added and polymerisation begins within each droplet, eventually forming a water—polymer slurry (or suspension), from which PS may easily be recovered, for example by centrifugation.

Emulsion polymerisation resembles suspension polymerisation in several ways. Both are heterophase processes in which the monomer and polymer are dispersed in water. In the emulsion process, however, polymerisation occurs not in the monomer disperse phase but in microscopic clusters or *micelles* of soap molecules in the aqueous phase. The micelles and the solid polymer particles to which they give rise are several orders of magnitude smaller than the particles formed in suspension polymerisation.

Suggestions for Reading

History, Technology and Economics

Allen, J. A., *Studies in Innovation in the Steel and Chemical Industries* (Manchester University Press, 1967).

Kaufman, M., *The History of PVC* (Applied Science, London, 1969).

McMillan, F., *The Chain-Straighteners* (Macmillan, London, 1979).

Morawetz, H., *The Origin and Growth of a Science* (Wiley, New York, 1985).

Morawetz, H., 'History of polymer science', in *Encyclopaedia of Polymer Science and Engineering*, 2nd edn, vol. 7, pp. 722–745 (Wiley, New York, 1987).

National Economic Development Office, *The Plastics Industry and its Prospects* (HMSO, London, 1972).

Reuben, B. G. and Burstall, M. L., *The Chemical Economy* (Longmans, London, 1973).

Seymour, R. B. and Kirshenbaum, G. S. (Eds), *High Performance Polymers: their Origin and Development* (Elsevier, New York, 1986).

Staudinger, H., *From Organic Chemistry to Macromolecules* (Interscience, New York, 1971).

Molecular Structure of Polymers

Billmeyer, F. W. Jr., *Textbook of Polymer Science*, 3rd edn (Wiley, New York, 1984).

Bovey, F. A., *Chain Structure and Conformation of Macromolecules* (Academic Press, New York, 1982).

Bovey, F. A. and Winslow, F. H., *Macromolecules: an Introduction to Polymer Science* (Academic Press, New York, 1979).

Brydson, J. A., *Plastics Materials*, 4th edn (Butterworth, London, 1982).

Sperling, L. H., *Introduction to Physical Polymer Science* (Wiley-Interscience, New York, 1986).

Naturally Occurring Polymeric Materials

Bolker, H. I., *Natural and Synthetic Polymers* (Dekker, New York, 1974).

MacGregor, E. A. and Greenwood, C. T., *Polymers in Nature* (Wiley, Chichester, 1980).

Wainwright, S. A., Biggs, W. D., Currey, J. D. and Gosline, J. M., *Mechanical Design in Organisms* (Edward Arnold, London, 1976).

Polymer Synthesis

Albright, L. F., *Processes for Major Addition-type Plastics and their Monomers* (McGraw-Hill, New York, 1974).

Allcock, H. R. and Lampe, F. W., *Contemporary Polymer Chemistry* (Prentice-Hall, Englewood Cliffs, NJ, 1981).

Flory, P. J., *Principles of Polymer Chemistry* (Cornell University Press, Ithaca, NY, 1953).

Goodman, M., and Falcetta, J. J., 'Polymerization', in H. S. Kaufman and J. J. Falcetta (Eds), *Introduction to Polymer Science and Technology* (SPE monograph no. 2) (Wiley, New York, 1977).

Heimenz, P. C., *Polymer Chemistry, the Basic Concepts* (Dekker, New York, 1984).

Odian, G. B., *Principles of Polymerization*, 2nd edn (Wiley, New York, 1981).

Saunders, K. J., *Organic Polymer Chemistry* (Chapman and Hall, London, 1973).

Seymour, R. B. and Carraher, C. E., Jr., *Polymer Chemistry: an Introduction* (Dekker, New York, 1981).

Stevens, M. P., *Polymer Chemistry: An Introduction* (Addison-Wesley, Reading, Mass., 1975).

Ziegler, K., 'Consequences and development of an invention', in Nobel Foundation, *Nobel Lectures: Chemistry, 1963–70*, pp. 6–24 (Elsevier, Amsterdam, 1972).

2

Morphology

Materials science seeks to explain the properties of materials in terms of structure. Chapter 1 dealt with the *molecular structure* of polymers, that is, with the internal organisation of the individual chain molecules. We next consider how the chain molecules are assembled to form the bulk material. In broad terms, in polymers as in other materials, we can envisage regular, ordered *crystalline* arrangements or irregular, random, *amorphous* arrangements of the constituent molecules. Furthermore, several levels of structural organisation may exist, and materials may reveal complex microstructures in the optical or electron microscope. It is to these observed forms and their inner structure that we refer when we speak of the *morphology* of polymeric materials. The properties of materials are governed by the subtle interplay of processes operating at all of these structural levels.

2.1 The Solid State: General Remarks

The multitude of individual atoms or molecules which comprise any solid material are held together by stabilising forces acting between them. These forces may be primary chemical bonds, as in metals and ionic solids, or they may be secondary van der Waals' forces in molecular solids such as ice, paraffin wax and in most polymers. The amount of stabilisation achieved by anchoring interactions between neighbouring molecules is at its greatest when molecules adopt regular rather than random arrangements. In the regular or crystalline arrangement (called the *crystal structure*) the molecules pack efficiently together to minimise the total intermolecular energy. Solids therefore tend to adopt such crystalline arrangements spontaneously. Crystal structures may be conveniently specified by describing the arrangement within the solid of a small representative group of atoms or molecules, called the *unit cell*. By multiplying identical unit cells in three directions the location of all molecules in the crystal is determined.

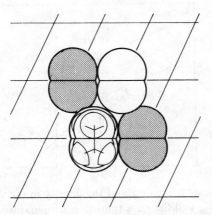

Figure 2.1   The crystal structure of solid ethane C_2H_6. The unit cell comprises
two molecules (unshaded)

A simple illustration of this representation of crystal structure is shown in
figure 2.1 for the non-polymeric hydrocarbon ethane — *see also* figure 1.3(b).
We shall see later in this chapter that regular crystalline arrangements of
molecules are found in many polymers.

Although the individual molecules are held firmly in place in the crystal lattice
they still possess vibrational and possibly rotational energy. Furthermore, the lattice
is never structurally perfect, and invariably contains a small proportion of positions
where molecules are missing or misplaced (vacancies and defects). The existence
of vacancies is important because it enables diffusional movement and reorganisation to occur in the solid state. As the temperature of the crystalline solid
approaches its melting point, the vibrational and diffusional motion becomes
more and more violent, finally causing the collapse of the crystal structure at
the melting point itself. In almost all materials, a small but sharp decrease in
density accompanies melting (figure 2.2) as the regular packing of the solid
gives place to the disorder of the liquid state. In the melt the forces acting
between the adjacent molecules are similar to those acting in the solid, but no
long-range order or structure persists, and all-pervasive translational diffusive
movement of molecules occurs.

Rising temperature ultimately destroys the crystalline order and causes melting.
However it is the reverse process that is of greater importance in understanding
the properties of solid materials. The morphology of the solid that is created on
cooling from the molten state is frozen into the material indefinitely and may
determine its critical physical properties from that time onwards. During the
solidification of molten material, as the temperature falls further below the
melting point, the ordered crystalline state attempts to emerge from the molecular disorder of the liquid.

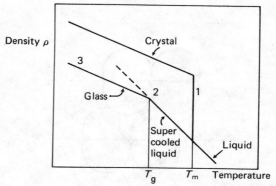

Figure 2.2   Temperature dependence of the density in the vicinity of the melting temperature T_m and the glass transition temperature T_g. There is a discontinuity 1 in the density ρ at the melting point of a crystalline solid. In the case of a liquid which supercools 2 to form a glass 3 T_g is marked by a change of slope in $\rho(T)$ but no discontinuity. Note that where both can be obtained, crystalline and glassy forms of the same substance have approximately the same volume expansivity $\rho[d(1/\rho)/dT]$

However, the regular arrangement of molecules cannot arise instantaneously, for it takes time for the molecules to find their places in the developing structural lattice. In fact it is impossible for the process of solidification to occur uniformly *at the molecular level* throughout the body of the melt. Crystalline regions start to grow at isolated points (*nuclei*). Once nucleated the embryonic crystallites grow by addition of further molecules, extending throughout the supercooled liquid or until they encounter the boundary of another growing crystalline region.

The crystallisation of a large amount of material on a single nucleus results in the formation of a *single crystal*. Well-formed individual single crystals can be obtained only under carefully controlled conditions (often more easily from solution than from the melt). Normally nucleation occurs at a number of points in the liquid and a *polycrystalline* solid is formed. The final size of the individual crystal grains or crystallites depends of course on the concentration of nucleating centres. The grain boundaries frequently play an important role in the properties of materials since they are lines of weakness and discontinuity. The grain pattern often dominates the microstructural appearance of materials such as metals and ceramics under the microscope, and crystalline regions occur in many polymeric materials.

The rate at which crystalline regions grow in a liquid below its melting point usually depends on the ease with which molecules can move about within the liquid. This mobility is largely controlled by the viscosity. Certain substances which have a high viscosity at the melting point are very reluctant to crystallise. If cooled quickly some way below the melting point the viscosity rises to such high values that crystallisation may be deferred indefinitely. Such substances at still lower temperatures show most of the properties of solids, but completely lack crystallinity. They are said to be *amorphous* and show the properties characteristic of glasses. Such substances do not show a sharp *phase change* from solid to liquid at a definite melting point, but alter gradually from recognisable solids to recognisable liquids over a range of temperature (figure 2.2). In the transition region these materials show interesting and important rheological properties. The lower limit of the transition range is marked by a change in, among other things, the coefficient of expansion (although not a discontinuity in density) which defines a characteristic *glass transition temperature* T_g. T_g should not be regarded as the counterpart of the melting point T_m, since substances which can be obtained both as crystalline and amorphous solids according to conditions of preparation generally have $T_g = \frac{2}{3} T_m$, approximately. T_m marks the upper limit of the viscoelastic region (*see* chapter 3 for a development of these ideas). Inorganic glasses include the important technical glasses based on silica. A wide range of normally crystalline substances, both organic and inorganic, can be obtained in glassy forms by rapid quenching of the melt to temperatures well below the melting point. Organic glasses of major importance are formed by certain polymers.

Crystallinity is the normal condition of solid substances, not simply a peculiarity of exotic materials such as gemstone minerals. However, most crystalline substances reveal their crystallinity only in their microstructures. They are almost invariably polycrystalline and rarely exhibit any outward crystalline form or habit. Polymers with a regular primary molecular structure are no exception to this rule of materials science and frequently form solids of well-developed crystallinity. On the other hand, polymeric substances which are *structurally irregular* at the molecular level, such as randomly-branched thermo-sets, most random copolymers and most atactic linear polymers, cannot possibly form crystalline arrangements and are necessarily amorphous. In fact, of all the major classes of materials, the polymers most strikingly combine crystallinity and non-crystallinity, order and disorder, in the solid state.

2.2 Crystal Structure of Polyethylene

Much of our knowledge of the crystal structures of materials of all kinds has come from the study of how X-rays are diffracted as they pass through solid substances. X-ray diffraction has thrown light on many areas of both the physical and biological sciences and it is established as one of the most powerful of experimental techniques. From early work on simple ionic solids, the X-ray methods are now so highly developed that the structures of intricate biopolymers can be explored.

X-rays are electromagnetic radiations of wavelength of the order of 100 pm. The wavelength is thus similar to the distance between atoms in solids, and consequently the X-radiation passing through solid substances is strongly scattered (diffracted) by the lattice of atoms or molecules. A narrow beam of X-rays passing through a crystalline solid is split into a number of distinct secondary beams, the direction and intensity of which are characteristic of the material, and in principle allow the location of atoms in the unit cell to be found. If the solid under study is amorphous rather than crystalline discrete secondary beams are absent, and the X-radiation is scattered diffusely. This reflects the absence of long range atomic regularity in the amorphous solid (figure 2.3).

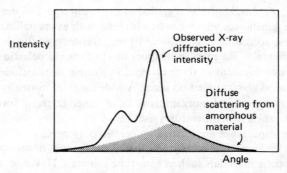

Figure 2.3   Low density polyethylene LDPE: X-ray diffraction intensity plotted against diffraction angle. The shaded portion is the contribution of amorphous material. Analysis of X-ray data provides an important method of determining the degree of crystallinity of semicrystalline polymers

Some of the earliest specimens of polyethylene were subjected to X-ray study in 1933, which showed at once a diffraction pattern very similar to that of low molar mass paraffin wax. The crystalline features of the PE diffraction pattern showed that the solid material contained *regular* arrangements of $-CH_2-$ chains side by side. In the individual molecular chains the C atoms are coplanar (figure 2.4). Thus the skeleton of C—C bonds had a highly regular zigzag conformation. In this coplanar conformation the separation of any two non-adjacent C atoms on the chain has the greatest value that can be achieved by bond rotation. It is, therefore, a *fully extended* conformation (*see* p.10). Crystallisation is accompanied by a considerable decrease in entropy but this is offset by an increase in cohesion because adjacent molecules are now arranged in the most favourable relation throughout the unit cell. Other common crystalline polymers adopt either this fully extended planar zigzag or a helical conformation which allows more room for bulky side groups (figure 2.4).

Two significant differences between the paraffin wax and polyethylene X-ray diffraction patterns were apparent. A slight blurring of the crystalline

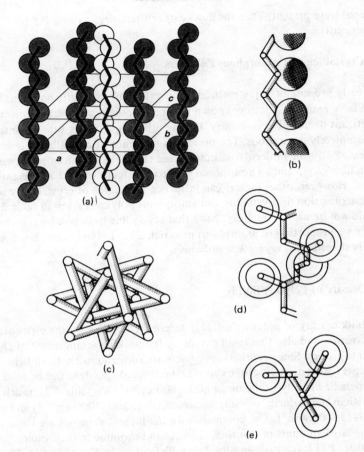

Figure 2.4   (a) Crystalline packing in polyethylene PE, showing the unit cell.
a, b and *c* define the crystallographic axes. Individual chains adopt a planar
zigzag conformation. (b) — (e) Chain conformations in crystalline polymers:
(b) syndiotactic poly(vinyl chloride)PVC, side view showing planar zigzag carbon
chain with bulky chlorine atoms alternately above and below the plane of the
chain; (c) polyisobutylene PIB (crystallised by stretching), end view showing the
regular helix formed by the carbon chain atoms; (d) isotactic polypropylene PP,
side view; and (e) end view, showing triangular helix

features indicated that the crystalline regions were quite small, perhaps 10–20 nm
across. Further, there was a featureless, smeared background of X-ray intensity
which clearly showed the existence of amorphous regions. In these early high
pressure polyethylenes roughly equal amounts of crystalline and amorphous

material were present. Thus the *degree of crystallinity* is about 0.5 (or 50 per cent).

2.3 Crystalline and Amorphous Polymers

The early promise of X-ray methods in investigating synthetic polymers has been fully realised. It is now known that many polymer materials possess a significant degree of crystallinity. On the other hand, many others are found to be completely amorphous. The presence or absence of crystallinity is determined *principally* by the molecular structure, and thus the division of polymers into a crystalline group and an amorphous group is both convenient and scientifically sound. However, other factors can influence the degree of crystallinity so that the classification of crystalline and amorphous polymers given in table 2.1 should not be used too rigidly. Note that crystalline bulk polymers invariably contain a proportion of amorphous material. Such polymers are, however, widely described as *crystalline polymers*.

2.4 Density of Polymer Solids

The bulk density of polymer solids is determined mainly by the elemental composition and also to a small extent by the packing arrangement of chains and side-groups. Since synthetic polymers are formed mostly from light elements (carbon, hydrogen, oxygen and nitrogen), the densities of solid polymers lie broadly in the range 800–1800 kg/m^3 (*see* table 2.2), much lower than the majority of inorganic materials (2200–4000 kg/m^3) and metals (2700–11500 kg/m^3). The polymers with the highest densities are those containing large amounts of fluorine, chlorine and bromine in their molecular structure: PTFE (relative density 2.28), PVDF (1.77), PVDC (1.65). The low density materials are hydrocarbon polymers with relatively open-packed structures: for example, polypropylene (0.90), ethylene–propylene copolymer elastomer EPM (0.86), and polymethylpentene (0.83). A rather insignificant thermoplastic, polybutylene, has the lowest density of all commercial polymers, 0.60. The bulk density of polymer materials may of course be greatly increased by compounding with heavy inorganic or metallic fillers (*see* chapter 6).

In semi-crystalline polymers, the measured density of a sample provides a simple way to estimate the degree of crystallinity. The density of the crystalline material may be calculated from the dimensions of the unit cell; the density of the amorphous form may be estimated from the melt density (*see* figure 2.2), using the thermal expansivity to extrapolate data to lower temperatures. For example, in the case of polyethylene, the crystal relative density is 1.000 and the relative density of the amorphous form is commonly taken to be 0.855. The degree of crystallinity *dc* of a polyethylene sample of density ρ may therefore be simply estimated from the relation $dc = (\rho - 0.855)/(1.000 - 0.855)$.

TABLE 2.1
Crystalline and amorphous polymers

Crystalline polymers	Amorphous polymers
General characteristics	
stereoregular homopolymer with strong interchain forces, *or* certain block copolymers	atactic homopolymer, *or* random copolymer
LDPE	Chlorinated PE
HDPE	EPDM and other ethylene
PP	copolymers
PTFE	PS (atactic commercial)
PA	ABS
Cellulose	PMMA
Polymethylpentene	AU
PETP, PBTP	EU
PC	Thermosets:
PEO	UF
POM	PF
PEEK	MF
	EP
	UP
	PUR

borderline cases

$$\left.\begin{array}{l} \text{NR} \\ \text{PIB} \\ \text{IIR} \\ \text{PVAL} \end{array}\right\} \begin{array}{c} \text{crystallise} \\ \text{at} \\ \text{high strains} \end{array}$$

PCTFE
PVC

2.5 Observed Microstructures: Spherulites

Crystalline polymers conceal their crystallinity very effectively — a polymer such as PP hardly fits the popular idea of a crystalline material. In fact, the hidden crystallinity of many polymers can be revealed quite easily, using the standard

TABLE 2.2
The relative densities of pure (uncompounded) polymer materials

Polybutylene	0.60
Polymethylpentene	0.83
Ethylene-propylene copolymer EPM	0.86
Polypropylene PP	0.90-0.92
Polyethylene PE	
LDPE	0.91-0.93
LLDPE	0.91-0.94
HDPE	0.96-0.97
Polybutene	0.91-0.92
Natural rubber	0.91
Butyl rubber IIR	0.92
Styrene-butadiene rubber SBR	0.93
Polyamide	
PA-12	1.02
PA-11	1.04
Acrylonitrile-styrene-butadiene copolymer ABS	1.04-1.07
Polystyrene PS	1.05
Poly(phenylene oxide)	1.06
Polyamide	
PA-6	1.12-1.13
PA-66	1.13-1.15
Polyacrylonitrile PAN	1.17
Poly(vinyl acetate) PVAC	1.19
Poly(methyl methacrylate) PMMA	1.19
Polycarbonate PC	1.2
Polyarylate	1.21
Polychloroprene rubber CR	1.23
Polysulphone	1.24
Polyetheretherketone PEEK	1.27
Poly(ethylene terephthalate) PETP	1.34-1.39
Poly(vinyl chloride)	1.37-1.39
Polyoxymethylene POM	1.41-1.43
Polyimide	1.43
Celluloses	1.58-1.63
Poly(vinylidene chloride) PVDC	1.65-1.70
Poly(vinylidene fluoride) PVDF	1.75-1.78
Polytetrafluorethylene	2.28

Notes In all cases the data refer to the uncompounded or unvulcanised pure polymer. In crystalline polymers where a density range is given, the density depends on the degree of crystallinity.

tool of the mineralogist, the polarising microscope. It is clear from figure 2.4 that a crystal should have highly developed directional features. The crystallographic axes define the principal directions in the structure geometry. Physical properties (for example, elastic modulus, electrical resistance or refractive index) may have different values along each of the axes, and such materials are generally *anisotropic*. Amorphous materials in contrast have the long-range disorder of liquids, lack any internal skeleton of characteristic axes, and are described as *isotropic*.

A small bead of PE melted, compressed into a thin film between microscope slide and coverslip, and allowed to cool through the melting point rapidly, shows little when viewed in unpolarised, transmitted light. The change in viewing the film between crossed polarisers is striking. An intricate microstructure is visible, the most prominent features of which are roughly circular structures known as *spherulites*, which may be as large as a millimetre across. Spherulites are sometimes seen in the crystalline microstructures of minerals and their appearance in any polymer is direct evidence of crystallinity. Figure 2.5(a) shows the appearance of spherulites in PE observed in polarised light. Two features of the individual spherulite are immediately evident. First, the Maltese cross (four symmetrically disposed sectors of high extinction) attests to the presence of birefringent units arranged with overall spherical (strictly, circular) symmetry. Second, the pattern of concentric circles reveals the existence of another level of long-range microstructural regularity. Although spherulites are normally studied in thin polymer films, it is generally accepted that spherical spherulites are the normal microstructural units formed in crystalline bulk polymers.

2.6 Morphology of Polymer Materials

Clearly, spherulites are high in the structural hierarchy — the crystalline regions revealed by X-ray diffraction being of dimensions too small for detection in the optical microscope, and several orders of magnitude smaller than the spherulite. The precise organisation of microstructural features in crystalline polymers remains the subject of active investigation and is not completely clear, but the following description is adequate for present purposes.

X-ray evidence showed that at the molecular level the polymer chains pack together side by side in the crystalline regions. The dimensions of the crystalline regions, however, are much smaller than the length of the polymer chains. What is the relation between the crystalline regions and the amorphous regions?

In the 1950s it was discovered that by crystallisation from very dilute solution single crystals of certain polymers could be obtained. These minute crystals (visible in the electron microscope) invariably took the form of very thin plates or *lamellae* — figure 2.5(b). X-ray study of polymer lamellae uncovered an unexpected fact: the polymer chain direction lies *transverse* to the plane of the lamella. It was then proposed that the polymer chains fold at regular intervals along the chain in the manner shown in figure 2.6(a). The lamellar thickness (hence the chain-fold interval) is extremely consistent, although it varies with the crystallisation conditions, such as temperature and solvent.

What happens to the molecule at the fold surface is probably of particular significance. Figure 2.6(a) is idealised, and we can think of several other states which molecules could adopt: long 'switchboard' loops, immediate re-entry to the crystal face, free tails not incorporated into the crystalline unit — figure 2.6(b), (c). In crystallisation from dilute solution, there is evidence that the chains fold rather tidily with little disorder at the fold surface. 'Adjacent re-entry', whilst it occurs, is not dominant: suggesting a model combining features of figure 2.6(a) and (b). In melt crystallisation, the re-entry is more random, some molecules no doubt acting as ties between different crystallites.

Studies of lamellar crystals precipitated from dilute solution have produced a rich harvest of unexpected findings of great intrinsic interest. But the conditions under which such single crystals are prepared are very different from those attending the solidification of polymer materials in technological processes. We approach, therefore, a central question: what is the relation between the

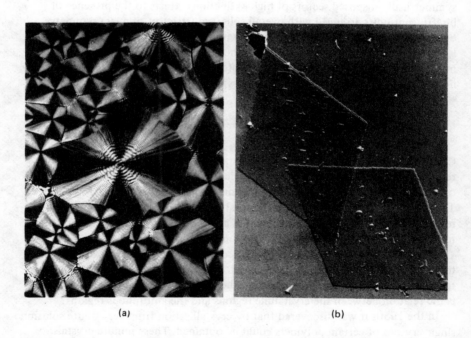

(a)   (b)

Figure 2.5   (a) Polyethylene observed in transmitted polarised light, showing the spherulitic morphology. (b) Two polyethylene single crystals (electron micrograph x 6750)—courtesy of V. F. Holland, Monsanto

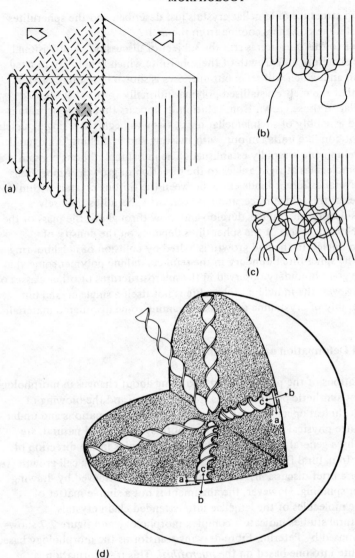

Figure 2.6 (a) Model of part of a lamellar polyethylene crystal with a regular folded chain structure. The unit cell, *see* figure 2.4(a), is shown shaded. The vertical lines represent the axes of the individual molecular chains. The loops at the fold surface depict orderly chain folding. (b), (c) Alternative models of the lamellar polymer crystal with disordered material at the fold surface, including switchboard loops and loose ends. (d) Structural organisation within the spherulite showing orientation of crystallographic axes [reproduced by permission from P. J. Barham and A. Keller, *J. Materials Sci.*, **12** (1977) 2141–2148]

formation and structure of lamellar crystals just described and the spherulites found in crystalline polymers cooling from the melt?

The answer to this question is still the subject of discussion. A provisional model of the structure and growth of the spherulite which is unlikely to need major revision and which will serve our purposes is shown in figure 2.6(d). It does appear that the melt crystallised polymer normally forms chain-folded lamellae similar to those grown from solution. It appears that the spherulite is a complicated assembly of such lamellar units. Growth begins at a single nucleus, from which ribbon-like units fan out, with twisting and branching. An overall spherical development is rapidly established. The individual ribbons contain chain-folded molecules at right angles to the direction of growth. Amorphous material of several different kinds exists between the fibrous crystals within the spherulites at the fold surface, and between the spherulites themselves. During crystallisation spherulites develop and grow throughout the mass of the polymer. The ultimate size of the spherulites depends on the density of crystallisation nuclei. In the end growth is halted by collision of neighbouring spherulites. The spherulite boundary in the semi-crystalline polymer somewhat resembles the grain boundary observed in the microstructures of other classes of materials. However, the individual spherulite is not itself a single crystal but rather an assembly of crystalline growths containing some disordered material.

2.7 Effects of Deformation on Morphology

Gross deformations of the polymer solid also bring about changes in morphology. The drawing of synthetic polymer fibres (*see* chapter 6) and the blowing of film are two technical operations involving large strain deformations and under-taken to enhance physical properties. Fibres, both synthetic and natural, are characterised by a general alignment of the polymer chains in the direction of the fibre axis. In natural fibres this orientation is achieved during cell growth (*see* section 6.9 for a brief discussion). In synthetic fibres it is achieved by drawing the fibres after spinning. However, the alignment is not a simple matter of pulling out the molecules of the lamellae into extended chain crystals.

Microstructural studies indicate a complex morphology and figure 2.7 shows the model proposed by Peterlin. Fibre-drawing transforms the morphology based on the spherulite into one based on the *microfibril*. This transformation is initiated by the fracture of lamellae under the action of the drawing stress. The chain molecules of the lamellae remain intact and break away from the lamellar fracture surface as a succession of small blocks within which the folded chain structure is preserved. As the drawing proceeds these blocks align to form the basic microfibril unit of the fibre. The blocks lie transverse to the axis of the microfibril (and the draw direction), firmly held together by tie molecules. The microfibril is strong and highly oriented at the molecular level. It is about 10–20 nm thick, but perhaps 10 μm long. Bundles of microfibrils produced by the failure of adjacent lamellae form *fibrils*, held together by van der Waals' forces and some of the original interlamellar tie molecules. The mechanical properties of drawn fibres depend on the deformation and slippage of microfibrils within the fibril.

2.8 Crystallisation

The theory of polymer crystallisation is concerned with the question raised earlier in this chapter: how does polymer morphology arise during solidification? More explicitly, how does the growth of crystallites and spherulites depend on temperature, rate of cooling and on the molecular properties of the individual polymer? This is profoundly important for polymer processing technology since the properties of the polymer material are strongly influenced by the morphology. By controlling crystallisation we achieve some control of morphology and therefore of material properties.

All crystalline regions grow by accretion of material from the melt on to nuclei. Homogeneous nucleation is the process by which nuclei are formed spontaneously throughout the melt (or solution) by random encounters of small numbers of molecules. Such *sporadic* nucleation becomes more probable the further the temperature falls below the melting temperature T_m. There is some evidence that in many polymers crystallisation of a *remelted* solid is nucleated by small crystals which have persisted from the initial solid. In a

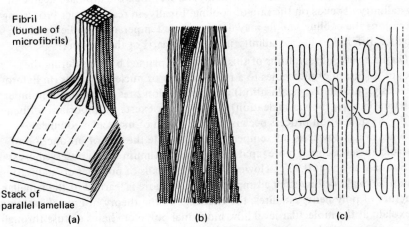

Fibril
(bundle of
microfibrils)

Stack of
parallel lamellae

(a)   (b)   (c)

Figure 2.7   (a) Transformation of spherulite lamellae into microfibrils.
(b) Fibrillar model of the drawn polymer fibre. (c) Molecular organisation in the fibril, showing microfibrils, crystal blocks and tie molecules. (After Peterlin, A., 'Mechanisms of deformation in polymeric solids', *Polymeric Materials*, American Society for Metals, 1975, p. 208)

polymer with a wide molar mass distribution melting may occur over a range of some degrees, and at temperatures above the so-called melting point unmelted crystallites may persist to act as nuclei. Such an explanation seems necessary to explain the observed dependence of crystallisation on thermal history. In any case, homogeneous nucleation is markedly affected by crystallisation temperature.

Heterogeneous nucleation occurs when extraneous solid particles or surfaces initiate crystal growth. The number of heterogeneous nucleation sites is not much altered as the melt cools. Little is known about the mechanisms of heterogeneous nucleation but it is of some technological importance, since it may occur on the walls of a container (such as a mould) or on the surface of solids such as pigment dispersed in the polymer. In some polymers, *nucleating agents* are deliberately added to the polymer to control the number density of spherulites and hence the final spherulite size. This is done for example in the injection moulding of polyamides such as PA-6. Finely dispersed silica (or indeed small quantities of high melting temperature polyamides) provide a controlled nucleation and generate a fine spherulitic texture.

The rate of crystallisation can readily be determined by measuring either the rate of change of the sample density or rate of growth of spherulites under the microscope. It is commonly found that the rate of crystallisation passes through a maximum at a temperature about $0.1T_m$ below the melting point. If the number of nuclei is constant (as may be the case in heterogeneous nucleation) then the spherulite size is not affected by the growth rate. On the other hand, if nuclei are forming continuously the spherulite size is determined by the ratio of nucleation rate and growth rate. If the crystallising polymer is cooling steadily (rather than being held at a constant temperature) then the ultimate degree of crystallinity depends on the rate of cooling. Finally, in real polymer processing operations the cooling sample may develop large temperature gradients which produce differences in crystallinity in different parts of the product.

Qualitatively, observations of this kind are explained by considering the rates of diffusion of molecules to a fixed number of nucleation sites (in heterogeneously nucleated crystallisation) or to nucleation sites increasing in number with time (homogeneous nucleation). Temperature exerts an important influence on the crystallisation process because the diffusion of molecules to nuclei becomes much slower as the temperature falls while the rate of production of sporadic nuclei becomes more rapid: hence the maximum in the observed crystallisation rate at about $0.9T_m$. However the fine details of polymer crystallisation cannot be explained by such a simple model. As more is learned of the lamellar structure of polymer crystallites, the challenge to the theory of crystallisation is to explain at the molecular level how individual polymer chains diffuse through the melt to attach themselves to the exposed crystal face. It appears that the attachment of each chain is itself a kind of nucleation, which once achieved is followed by deposition of the rest of the chain-folded polymer molecule on the crystal face. Exactly how this occurs is of course very closely related to the issue of chain re-entry discussed in section 2.3 above. Present evidence suggests that at temperatures not far below T_m, chains attach themselves to growing crystal faces rather infrequently and deposit in an orderly way, achieving adjacent re-entry with little mutual interference between separate molecules. At lower temperatures, the secondary nucleation is rapid and and frequent; as a result, in the frantic competition for lattice sites, a rather untidy crystallisation occurs, probably with little adjacent re-entry and with switchboard loops and loose ends of the kind shown in figure 2.6(b) and (c).

Nucleation is thus a process of creating local order among small numbers of molecular units in neighbouring chains. We have already mentioned the parts played in this both by small crystals which persist in the melt and by extraneous solid particles, but in a number of important practical processes this essential ordering step is promoted by subjecting the amorphous polymer to mechanical stresses which orient the molecular chains to a certain degree. Specific examples of such processes are those for nylon and polyester textile fibres and polyester film. In the latter case orientation and subsequent crystallisation are induced in two directions (biaxial orientation) by a controlled sequence of heating and stressing. If the rate at which local order produced under typical applied stresses is compared with that occurring by chance in unstrained material (thermal crystallisation) increases in the nucleation rate of the order of 10^6 upwards are predicted. This is broadly commensurate with the rates of crystallisation observed in practice in the two situations. Crystallisation from melts under stress gives rise to a fibrillar rather than a spherulitic morphology, in which crystalline fibrils develop along the direction of flow.

2.9 Molecular Motion in Polymers

In chapter 1 we described the relatively free rotation which may occur about single bonds in the molecular structure. We noted the resultant flexibility of the linear polymer chain in the melt, in solution and in the viscoelastic state. The present chapter has discussed the processes of crystallisation, which involve the translational diffusion of entire molecules and cooperative changes of conformation among assemblies of many molecules. We conclude this chapter by drawing together the threads of our account of molecular motion in polymers.

The nature and the extent of molecular mobility are controlled by the temperature. At the very lowest temperatures, close to absolute zero (0 K), the polymeric solid possesses so little thermal energy that all atomic positions are stationary. Only a slight residual fuzziness remains, arising from the quantum zero-point vibrational energy. As the temperature rises, the thermal energy acquired is partitioned between all the various possible modes of motion. The different modes however require different amounts of energy to activate them and therefore there exist threshold temperatures for the unfreezing of each. Experimentally these threshold temperatures can be detected as transitions in a wide variety of physical parameters (*see* figure 2.8). In amorphous polymers

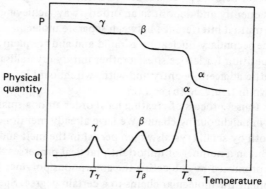

Figure 2.8   The occurrence of various types of motion may be revealed in the temperature dependence of many properties. Properties measured by quantities of type P (for example, mechanical shear modulus, electrical permittivity, heat capacity, refractive index) show one or more transition zones $\alpha, \beta, \gamma \ldots$ roughly defining transition temperatures $T_\alpha, T_\beta, T_\gamma \ldots$. Properties measured by quantities of type Q (such as the mechanical and dielectric loss tangents) show one or more maxima (*see* text, chapters 3 and 4)

the glass transition is the principal or *primary* transition observed; we have already indicated how the associated temperature T_g may be determined from density measurements. In crystalline polymers the most prominent transition is associated with melting, defining T_m. In addition, a glass transition T_g is generally discernible, arising from the amorphous material present in semi-crystalline polymer solids. Some data are collected in table 2.3. The glass transition is commonly labelled α and secondary transitions $\beta, \gamma \ldots$ in order of decreasing transition temperature ($T_\alpha > T_\beta > T_\gamma \ldots$).

The main modes of motion which may in principle occur in polymer materials are illustrated in figure 2.9. To which of these motions do the various observed transitions correspond? This question cannot yet be fully answered. In crystalline polymers T_m marks the onset of translational diffusion of the molecular centre of mass: the collapse of the crystalline lattice and the transition to the molecular brownian motion characteristic of the liquid state. The nature of the glass transition remains somewhat unclear. The discontinuity in the expansivity at T_g suggests that the glass transition signals a change in the availability of space between the molecules. Such space (or *free volume*) must exist for large-scale motions of the polymer chain to occur. Above T_g these large-scale motions take the form of concerted movements of the backbones of neighbouring polymer chains, each segment comprising perhaps 10 or 20 mers. It is these segmental motions which permit the large deformations characteristic of the viscoelastic state. As the polymer cools to the glass transition the free volume

TABLE 2.3
Glass transition temperature T_g of common polymers

Polymer	T_g (°C)
PE	−90/−35
PP	−10
polymethylpentene	30
PS	95
PAN	105
PVC	85
PVF	−20/45
PVDC	−15
PA 6	50
PA 66	90
PA 610	40
PMMA	105
POM	−90/−10
poly(phenylene oxide)	210
PC	150
PETP	65
CA	105
NR	−75
CR	−45
NBR	−20

These are approximate values; where two temperatures are given the assignment of the glass transition remains doubtful. T_m is independent of chain length for high molar mass polymers, but *falls* somewhat as chains become very short.

diminishes to a critical value below which segmental motion is impossible or at least extremely slow. Below T_g there remain only the more limited motions such as the three-bond and five-bond crankshaft rotations of the main chain, and rotations of side groups or parts of side groups (*see* figure 2.9). Some of these persist down to very low temperatures indeed. A transition identified with methyl group rotation occurs at about 6 K (−267 °C) in PMMA and at about 40 K in PP. In crystalline polymers some secondary transitions are attributed to structural features of the crystal lattice, such as the motion of packing defects. In practice the assignment of observed secondary transitions to particular motions is frequently rather uncertain and tentative.

Recently, the motion of chain molecules in polymer melts has been studied intensively as a problem in molecular statistical physics. The results are significant for polymer materials science because they help us to understand two important

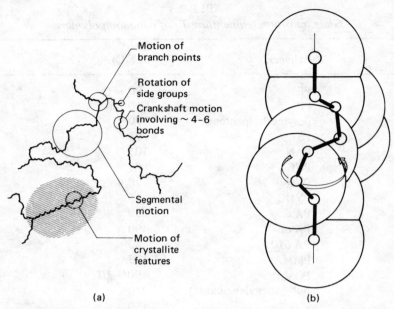

Figure 2.9 (a) Various types of molecular and microstructural motion.
(b) Crankshaft motion in a linear chain, illustrated by a short segment of a
polyethylene chain comprising eigth backbone carbon atoms. A crankshaft
rotation of the four inner C atoms is possible even if the four outer atoms are
stationary. Similar crankshaft motions involving three bonds are possible

practical matters: the fluid mechanics of polymers in melt processing operations
such as extrusion and injection moulding; and the diffusion of molecules during
crystallisation.

Experimentally it is well established that the viscosity of a polymer melt consisting of linear chains increases rapidly with the chain length. For linear
polymers, the viscosity varies as $M^{3.4}$ above a critical chain length (figure 2.10).
This reflects the fact that relative motion of the chains must depend on the rate
at which chains can disentangle themselves, and that this becomes rapidly more
difficult as the chain length increases. (Below the critical chain length the viscosity varies only as $M^{1.7}$). Some progress has been made by considering the
motion of a single polymer molecule wriggling along a tortuous path formed by
its neighbours. This *reptation* concept, associated with P.-G. de Gennes, has been
developed by Doi and Edwards into a fully quantitative theory of polymer chain
dynamics.

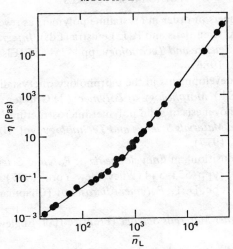

Figure 2.10 Viscosity of a polymer melt: dependence on molar mass or chain length. For linear polymers, the critical chain length for entanglements is about 500 bonds. [Polyisobutylene data]

Suggestions for Reading

Bassett, D. C., *Principles of Polymer Morphology* (Cambridge University Press, 1981).

Billmeyer, F. W., Jr., *Textbook of Polymer Science*, 3rd edn (Wiley, New York, 1984).

Bird, R. B., Armstrong, R. C. and Hassager, O., *Dynamics of Polymeric Liquids*, 2nd edn, 2 vols (Wiley, New York, 1979).

Clark, E. S., 'Structure of crystalline polymers', in *Polymeric Materials*, ch. 1 (American Society of Metals, Metals Park, Ohio, 1975).

de Gennes, P.-G., *Scaling Concepts in Polymer Physics* (Cornell University Press, Ithaca, NY, 1979).

Doi, M. and Edwards, S. F., *The Theory of Polymer Dynamics* (Clarendon, Oxford, 1986).

Hemsley, D. A., *Light Microscopy of Synthetic Polymers* (Oxford University Press, 1984).

Ingram, P. and Peterlin, A., 'Morphology', in *Encyclopaedia of Polymer Science and Technology*, vol. 9, pp. 204–74 (Wiley, New York, 1968).

Keller, A., 'The many faces of order in solid polymers', *Plastics and Polymers*, **43**(163) (1975) 15–29.

Keller, A., 'On the methodology of morphological and structure research in solid polymers', in F. Ciardelli and P. Giusti (Eds), *Structural Order in Polymers*, pp. 135–179 (Pergamon, Oxford, 1981).

Keller, A., 'Some facets of order in crystalline polymers as revealed by polyethylene' in L. A. Kleintjens and P. J. Lemstra (Eds), *Integration of Fundamental Polymer Science and Technology*, pp. 425–455 (Elsevier Applied
Science, London, 1986).

Keller, A., 'Recent developments in the morphology of crystalline polymers',
in B. Sedlacek (Ed.), *Morphology of Polymers* (de Gruyter, Berlin, 1986).

Magill, J. H., 'Morphogenesis of solid polymer microstructures', in J. M. Schultz
(Ed.), *Treatise on Materials Science and Technology*, vol. 10A (Academic
Press, New York, 1977).

Roe, R.-J., 'Glass transition', in *Encyclopaedia of Polymer Science and Engineering*, 2nd edn, vol. 7, pp. 531–544 (Wiley, New York, 1987).

Sawyer, L. C. and Grubb, D. T., *Polymer Microscopy* (Chapman and Hall,
London, 1987).

Schultz, J. M., *Polymer Materials Science* (Prentice-Hall, Englewood Cliffs, NJ,
1974).

Sharples, A., 'Crystallinity', in A. D. Jenkins (Ed.), *Polymer Science*, ch. 4
(North-Holland, Amsterdam, 1972).

Sperling, L. H., *Introduction to Physical Polymer Science* (Wiley–Interscience,
New York, 1986).

Voigt-Martin, I. H. and Wendorff, J., 'Amorphous polymers', in *Encyclopaedia
of Polymer Science and Engineering*, vol. 1, pp. 789–842 (Wiley, New York,
1985).

Wunderlich, B., *Macromolecular Physics*, vol. 1 *Crystal Structure, Morphology,
Defects* (Academic Press, New York, 1973); vol. 2 *Crystal Nucleation, Growth,
Annealing* (Academic Press, New York, 1976); vol. 3 *Crystal Melting*
(Academic Press, New York, 1980).

Wyatt, O. H. and Dew-Hughes, D., *Metals, Ceramics and Polymers* (Cambridge
University Press, 1974).

Young, R. J., *Introduction to Polymers* (Chapman and Hall, London 1981).

3

Mechanical and Thermal Properties

Everyday experience shows that polymeric materials display a remarkably wide range of mechanical behaviour, spanning brittle solid, rubber, leathery plastic, and strong fibre. Moreover, it is often evident that the mechanical character of a solid polymer is altered greatly by changes of temperature as small as a few degrees. In chapter 2 we saw that an amorphous polymer such as PMMA is brittle below its glass transition temperature T_g. At higher temperatures it softens progressively, turning gradually without any obvious discontinuity of property into a viscous liquid as the temperature rises.

We shall now consider how to describe this diversity of mechanical behaviour, paying particular attention to the sensitivity of mechanical properties to changes of temperature. This is a special characteristic of polymeric materials as a class, and one with important consequences for polymer processing and design.

3.1 Deformations: Stress and Strain

We apply the word *brittle* to a solid which will not support large deformations, and which fractures without appreciable yielding. If a specimen of such a material is stressed to failure in a tensile testing apparatus, the characteristic stress–strain relationship ideally resembles the curve shown in figure 3.1(a). It is clear that Hooke's law is obeyed, at least approximately

$$\epsilon = D\sigma \tag{3.1}$$

(D is called the *compliance*, the 'spring constant'); it is more usual in engineering contexts to express equation 3.1 with σ as the dependent variable:

$$\sigma = E\epsilon \tag{3.2}$$

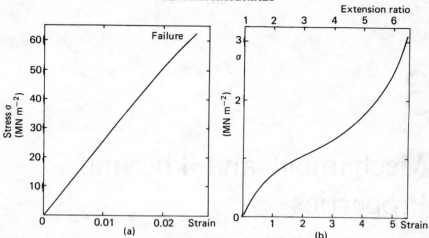

Figure 3.1 (a) Typical tensile stress–strain curve of a brittle polymer, such as PMMA or CA at a low temperature. (b) Typical tensile stress–strain curve for an unfilled rubber. Note that (a) and (b) have different stress and strain scales

where E is the tensile elastic modulus (Young's modulus). Such brittle behaviour in glassy polymers well below T_g is much the same in essence as the brittleness of many metals, inorganic glasses and ceramics. Failure occurs cleanly and suddenly, with no gross prefracture yield. At any strain below the failure strain the deformation is entirely reversible. Such behaviour is described as *elastic*. An elastic body spontaneously recovers its original state on removal of the applied stress, and the energy expended to produce the original deformation is totally recoverable.

3.2 Brittle Hookean Solids

Inability to support tensile strains greater than a few per cent indicates that there is little scope within the micro- or crystal structure for internal reorganisation of brittle solids in response to an imposed stress. The largeness of the elastic modulus reflects the large forces needed to move atoms even fractionally from their equilibrium positions by bond stretching, bond compression, or by a change of spacing between adjacent non-bonded atoms.

The calculation of the modulus depends on a detailed knowledge of the energies of interaction between all atoms in the solid. Although precise calculations may not be possible, it is interesting to note that for small strains there should be little difference between the behaviour in tension and compression. If the energy of interaction of a typical pair of neighbouring atoms is represented by the curve U (figure 3.2), then $d^2 U/dr^2$ in the vicinity $U = U_0$

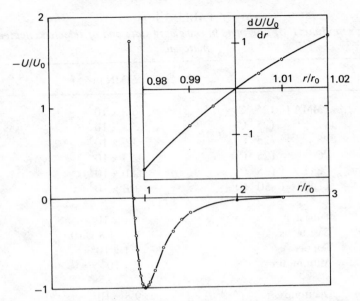

Figure 3.2 Energy of interaction $U(r)$ of non-bonded atoms or molecules calculated from the Mie equation $U = -a/r^m + b/r^n$ with $m = 6$ and $n = 12$ (van der Waals' interaction). U_0, r_0 are equilibrium energy and distance of separation. For many solid materials the bulk modulus K is proportional to $d^2 U/dr^2|_{r_0} = nmU_0/r_0^2$. Inset shows dU/dr for small strains $\pm 0.02 r_0$

represents the force/deformation ratio, which we may take as a rough measure of the modulus. For small deformations the force evidently depends approximately linearly on the deformation (thus Hooke's law) and is about the same in compression as in tension. In fact by Gruneisen's first rule the bulk modulus $K \approx nmU_0/9V_0$ where V_0 is the molar volume. If we consider the van der Waals' interaction between the CH_2 groups of neighbouring PE molecules we may take $U_0 = 8.0$ kJ/mol CH_2 and $V_0 = 22$ cm^3. Hence we obtain $K \approx 3 \times 10^3$ MN/m^2, in good agreement with the experimental value found for polycrystalline PE.

3.3 Elastic Moduli of Glassy Polymers

Table 3.1 lists the elastic moduli of a number of glassy polymers (at specified temperatures) and other engineering materials. The polymers are less stiff than structural metals and ceramics by one or two orders of magnitude. Where deformation involves primary bond stretching (as in metals, ionic solids, carbon fibre) moduli of $10^5 - 10^6$ MN/m^2 are found. Small strains in glassy polymers

TABLE 3.1

Young's modulus E of polymers in the glassy state and of selected engineering materials

	$E(\text{MN m}^{-2})$
PMMA (−125 °C)	6.3×10^3
(25 °C)	3.3×10^3
PS (25 °C)	3.5×10^3
PC (25 °C)	2.3×10^3
PETP (25 °C)	2.0×10^3
LDPE (−50 °C)	1.8×10^3
Tungsten	4.1×10^5
Cast irons	$1.2 - 1.8 \times 10^5$
Copper	1.1×10^5
Aluminium	7×10^4
Diamond	9.5×10^5
Dense aluminium oxide	4×10^5
High modulus carbon fibre	4.2×10^5
Soda glass	6.9×10^4
Fused quartz	7.3×10^4

apparently involve relative movements of non-bonded atoms (van der Waals' force fields), which is reflected in lower elastic moduli, $10^3 - 10^4$ MN/m^2.

3.4 Rubber Elasticity

We have described the stress—strain diagram obtained in a tensile test on a brittle solid. A similar test carried out on a rubbery polymer or *elastomer*, however, yields a very different stress—strain curve — figure 3.1(b). Once again the behaviour is elastic, since the specimen returns to its original state ($\sigma = 0$, $\epsilon = 0$) on removal of the load. None the less, three differences stand out. First, Hooke's law is not obeyed. In particular the modulus increases at large strains. Second, relatively enormous deformations may be attained. Rubbers may be repeatedly extended to as much as five times their unstrained length completely reversibly. This represents a strain perhaps one thousand times greater than the failure strain of many brittle solids. Third, the mean elastic modulus is relatively very low (two or three orders of magnitude less than that of a glassy polymer).

In rubbery solids, molecular reorganisation in response to applied stress is relatively easy, although ultimately restrained by a small concentration of cross-links between chains. (In the rubbery state of nonelastomers such as PMMA and PS, permanent chemical crosslinks are absent and chain entanglements serve to restrain the relative movement of chains. Such materials only exhibit full elastic recovery in short-term experiments, because the entanglements free themselves over a sufficiently long period of time permitting irreversible movements of chains in the solid.)

In the unstrained state (state A of figure 3.3) the molecular chains adopt the most probable random arrangements consistent with chemical crosslinking or entanglement coupling. To a good approximation, the sections of the molecules which lie between each pair of crosslinks behave as freely jointed chains, and the spatial distribution of these chains within the rubber network corresponds to the maximum in the gaussian function already shown in figure 1.6(b). The state A is not static at the molecular level because of the thermal motion of the chains; however, the distribution averaged over the very large number of conformational arrangements which can be adopted is constant. When the rubber is deformed (A → B) the separations of adjacent tie points in the network are changed in a manner which depends on the macroscopic strain. The deformation entails a shift in the random conformations of the freely jointed chains to less probable distributions. When the stress is removed the rubber rapidly and spontaneously recovers its unstrained condition as the chains return to a state indistinguishable

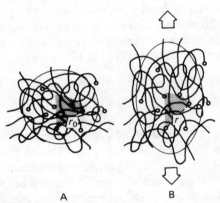

A   B

Figure 3.3   The rubbery state. A represents a small volume element of an unstrained material, B the same volume extended along the axis shown. Each line represents a chain segment between two tie points, depicted here as a free chain. The change $(r - r_0)$ in separation of adjacent tie points in each molecular chain depends on the angle θ (shown as a shaded sector). Averaged over the whole network $(r - r_0)$ is approximately proportional to the overall strain. (After Peterlin, A., 'Mechanisms of deformation in polymeric solids', *Polymeric Materials*, American Society of Metals, 1975, p. 181)

from A. The recovery from B to A occurs because random thermal motion tends to recreate the more probable distribution.

It is interesting to note that in compression (that is, bulk not uniaxial compression) the modulus of rubbery solids is much higher than in tension, similar in fact to that of other solids. This is so because the redistribution which occurs in extension and shear is no longer possible, and deformation can only occur by closer packing of molecules and compression of atom against atom.

This view of rubber elasticity provides the basis for a powerful statistical thermodynamic analysis of the rubbery state, which we may outline as follows. To transform state A to state B (figure 3.3) requires the expenditure of work because B has a higher free energy than A. We recall that the free energy is a composite quantity comprising enthalpy and entropy contributions. In elastomers, enthalpy differences between states like A and B are relatively small. Indeed an *ideal elastomer* is defined as one whose enthalpy is independent of strain (compare this with an ideal gas, whose enthalpy is independent of volume). In an ideal elastomer therefore the free energy difference between A and B (and hence the work necessary to deform the material) depends on the entropy difference between A and B, ΔS. ΔS can be calculated from the probability function $\omega(r)$ describing the distribution of chain conformations, since $S = $ constant $\times \ln \omega(r)$. In turn the work of deformation $= -T\Delta S$, where T is the absolute temperature. This suggests that the work required to attain a particular strain increases as the temperature increases. The implications that the tensile modulus of a rubber should increase with temperature and that a rubber in tension should contract on heating are both borne out by experiment. Another interesting result of this analysis, also confirmed experimentally, is that rubbers become hotter during adiabatic stretching and cool on retraction. Real rubbers approximate to a greater or less extent to ideal elastomers.

3.5 Viscoelasticity

We have now described two types of mechanical response in polymers (the responses of the glassy state and the rubbery state) which are approximately elastic. We have noted that amorphous polymers exhibit both types of behaviour (figure 3.4). The glassy state occurs at temperatures well below T_g and the rubbery state somewhat above T_g. Between the two is a *transition zone*, usually spanning some 20 °C, in which the stiffness decreases rapidly with increasing temperature. In lightly crosslinked polymers (elastomers such as vulcanised rubbers) the rubbery state persists up to the highest temperatures at which the material is thermally stable. In non-crosslinked amorphous polymers (such as PMMA and PS) the rubbery state extends only to the vicinity of another transition temperature (or terminal zone) above which the polymer melts and is essentially fluid.

The mechanical behaviour of crystalline polymers is also greatly influenced

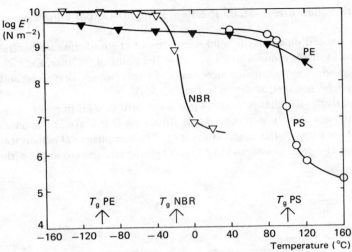

Figure 3.4   Variation of tensile modulus E' with temperature for a crystalline polymer HDPE and two amorphous polymers PS and NBR (an elastomer). The glass transition temperatures T_g are shown. PS is normally used at temperatures below its T_g and NBR above

by temperature (figure 3.4). The reduction in stiffness as the temperature rises is less pronounced than in amorphous polymers and the transition zone associated with T_g is generally broadened. The effect of crystallinity is to produce a leathery rather than a rubbery condition in the polymer above T_g. At still higher temperatures a reasonably well-defined melting temperature marks the appearance of the fluid state.

Figure 3.4 emphasises how sensitive to temperature is the mechanical behaviour of a polymeric material. At the lowest temperatures polymers, whether amorphous or crystalline, are brittle, approximately elastic, glasslike solids. At the highest temperatures uncrosslinked materials are molten, viscous substances, in which the response to an applied stress is flow. In the whole of the intermediate region, comprising the transition zone associated with T_g, the rubbery plateau, the leathery state of crystalline polymers, and the terminal zone, the mechanical behaviour shows features of both elastic solid and viscous liquid. Since the working temperatures of many engineering polymers fall in this intermediate region we must sometimes consider polymer solids as a class as *viscoelastic* rather than elastic bodies. The occurrence of viscous processes introduces time-dependence and sometimes irreversibility into the stress–strain behaviour of the polymeric materials. The central theme of polymer mechanics is the interrelationship of the variables stress, strain, time and temperature. In order to present a unified analysis of such varied mechanical behaviour it is useful to look again at the elastic response.

3.6 Elastic Solids: Stress–Strain Relations as Input–Output Functions

If the tensile test on an elastic solid is carried out at a uniform rate of straining, from zero to some terminal strain ϵ_1 below the point of fracture, the increments of strain produce simultaneous increments of stress (we say that stress and strain change *synchronously*), as shown in figure 3.5(a).

Alternatively, if a change in stress (say from zero to σ) is imposed instantaneously at t_1 (a step function of stress) the tensile strain response is immediate, ϵ rising to the constant value ϵ_1. The compliance D is therefore also a constant. If σ returns to zero at t_2 then ϵ returns also to zero at t_2, without any lag.

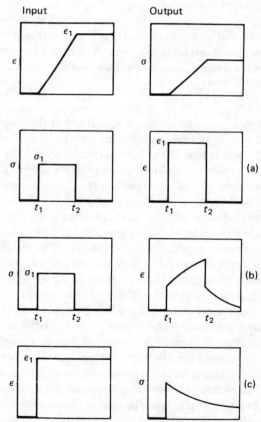

Figure 3.5   (a) Synchronous stress–strain behaviour in an elastic material.
(b) Creep, recovery and (c) stress relaxation in a viscoelastic material

3.7 Viscoelastic Materials

Amorphous polymers above T_g do not show this behaviour. For example, the response to a step function of stress may be of the general form shown in figure 3.5(b). Now D is no longer a constant. For the interval t_1 to t_2, σ has a constant value σ_1 but ϵ is a function of time $\epsilon(t)$, thus

$$D(t) = \epsilon(t)/\sigma_1 \qquad (3.3)$$

The form of figure 3.5(b) shows that the material continues to deform under constant load. This behaviour, known as *creep*, is in striking contrast to that of the elastic solid which deforms only while the load is being applied. Creep behaviour (to which we return in section 3.9) has some of the characteristics of liquid viscous flow, and time-dependent stress—strain relations of this general kind are thus described as *viscoelastic*. It is important to note that equation 3.3 describes *linear* time-dependent behaviour, i.e. ϵ and D are linearly related through the stress σ. If $D(t)$ is determined at one value of σ then the complete strain response can be calculated at any other stress. Such behaviour is often observed in amorphous polymers for small strains and short times. We return briefly in section 3.15 to the more general case of nonlinear viscoelasticity.

It is an important feature of linear viscoelasticity theory that the effects of a number of imposed stresses are additive. This is embodied in the *Boltzmann superposition principle*. It is of great practical importance since it means that the response of the material to a complicated stress (or strain) input function can be calculated from a quite small amount of data.

3.8 Stress Relaxation

An experiment to determine the creep response of a material imposes a stress (normally a constant load) and examines the time-dependent strain response of the specimen. If on the other hand a fixed strain ϵ_1 is imposed on the specimen, the measured stress is now time dependent — figure 3.5(c). The material exhibits a decreasing stress or *stress relaxation*, which for a linear viscoelastic body leads naturally to the definition of a stress relaxation modulus

$$E(t) = \sigma(t)/\epsilon_1 \qquad (3.4)$$

(compare this with equation 3.3).

Can we be more precise in describing how *viscoelasticity* combines the properties of the elastic solid and the viscous liquid? A very useful method of analysis lies in the construction (on paper only!) of ideal mechanical models which simulate the stress—strain—time behaviour of the real materials. The mechanical models are normally built up from only two kinds of unit: the hookean spring $\epsilon = D\sigma$ and the linear viscous element (usually represented by a dashpot, a perforated piston moving in a cylinder filled with a viscous fluid) for which

$d\epsilon/dt = \eta^{-1}\sigma$, where η is the viscosity. A series combination of spring and dashpot is known as a *Maxwell element* and a parallel combination is called a *Voigt element*. The response of the Maxwell element to a constant stress is

$$\epsilon = (D + t/\eta)\sigma \qquad (3.5)$$

ϵ increases linearly with time — figure 3.6(a). Similarly the response to a step function of strain is

$$\sigma = \epsilon \exp(-t/\eta D)/D \qquad (3.6)$$

Thus the stress relaxes exponentially as shown in figure 3.6(b); the rate of relaxation is determined by the characteristic time ηD. A long slow relaxation (with a long relaxation time) is caused either by a weak spring constant or a high viscosity. The underlying linearity of the system is revealed by the fact that the relaxation time depends only on D and η.

In the Voigt element the parallel combination prevents the development of creep at a constant rate under a constant load

$$\epsilon = \sigma D[1 - \exp(-t/\eta D)] \qquad (3.7)$$

As $t \to \infty$, $\epsilon \to \epsilon_\infty = \sigma D$

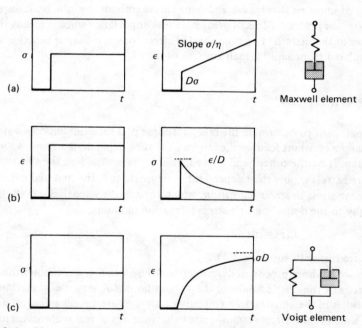

Figure 3.6 Maxwell and Voigt elements. (a) Maxwell element: step function of stress and strain response. (b) Maxwell element: step function of strain and stress response. (c) Voigt element: step function of stress and strain response

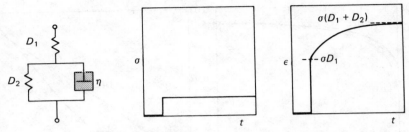

Figure 3.7 Creep compliance of a three-element model. Strain response to a step function of stress of spring, series-coupled to a Voigt element (a form of the standard linear solid)

This ultimate strain ϵ_∞ is the same as the strain developed in an elastic material of compliance D in the absence of a dashpot. The essential difference is that the strain response is not instantaneous but *retarded*. The quantity ηD is here the characteristic *retardation time* of the system. (To impose a step change of strain on a Voigt element is unrealistic as it implies an infinite instantaneous stress in the dashpot.)

More complex responses can be obtained by adding further units in series or parallel with Maxwell or Voigt elements. An important three-element model is shown in figure 3.7.

It is important to remember that such model descriptions of viscoelastic behaviour are highly simplified representations of the mechanics of polymer solids. Nevertheless there is some physical justification for combining reversible, elastic and irreversible, dissipative elements in this way. The merit of these models is in providing a basis for developing differential equations for $\sigma(t)$ and $\epsilon(t)$, which may later be generalised.

3.9 Creep and Stress Relaxation Experiments

The step input functions of stress and strain which we have just examined are of special interest because they correspond to two of the principal experimental arrangements in common use in testing the mechanical properties of plastics. These are the creep test and the stress relaxation test in uniaxial tension. Creep tests may also be carried out in flexure and in torsion. The response to a constant rate of strain may also be conveniently studied with conventional tensile testing machines. The response of the Maxwell, Voigt or three-element model to a constant imposed strain rate can readily be obtained.

Creep tests are particularly simple in concept, although careful design of apparatus is essential if high accuracy is to be attained in tests which may last for many months. The deformation of the specimen has been measured by a

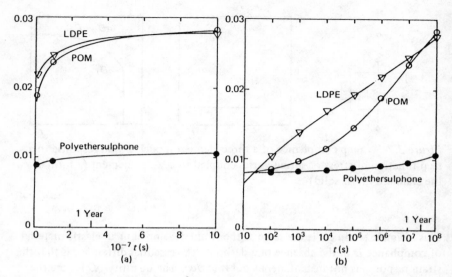

Figure 3.8 Tensile creep at 20 °C of LDPE (20 MN m^{-2}), POM copolymer (20 MN m^{-2}) and polyethersulphone (2 MN m^{-2}) (a) Strain plotted against time. (b) Strain plotted against log(time). (Data from ICI Ltd)

variety of mechanical, electromechanical and optical devices. Figure 3.8 shows the strain–time curves for three thermoplastics. It is clear that the behaviour of the thermoplastics somewhat resembles that of a Maxwell element: an instantaneous deformation followed by creep. However, if the creep strain is followed for long times it is frequently found to increase at a reducing rate. The creep rate of the Maxwell element on the other hand is constant and to that extent the Maxwell element is too simple a model. The three-element model is a significant improvement and shows an exponentially falling creep rate; moreover, complete strain recovery eventually follows removal of stress. The typical rubbery solid shows a retarded elastic response similar to that of a Voigt element. The strain–time curve of the Voigt element under constant stress is a simple exponential. This is not the case for real rubbers.

3.10 Empirical Creep Equations

Our discussion of the creep response of the Maxwell three-element model suggests that it should be possible to split the observed creep strains of polymeric materials into two parts

$$\epsilon = \epsilon_1 + \epsilon_2(t) \tag{3.8}$$

Here ϵ_1 is the instantaneous elastic strain and ϵ_2 is a time-dependent strain.

Although ϵ_2 does not in fact depend linearly or exponentially or in any other simple way on t it is often possible to find an *empirical* function which fits the observed data well

$$\epsilon - \epsilon_1 = \sigma f(t) \tag{3.9}$$

where σ is the stress. Knowledge of $f(t)$ then permits computation of ϵ at any time and any stress. Conversely a large body of creep data may be condensed into a single function f. For example, the creep of several thermoplastics at low stress may be represented approximately by the empirical function

$$\epsilon = \sigma(B + At^\alpha)$$

or

$$(\epsilon - \sigma B)/\sigma = At^\alpha = f(t) \tag{3.10}$$

For such materials the creep curve is usefully displayed on a logarithmic or semi-logarithmic plot (*see* figure 3.8(b), LDPE and polyethersulphone); at corresponding times the creep strain is proportional to the stress (linear behaviour).

3.11 Dynamic Response

We have seen how the characteristic relaxation time or retardation time arises in response to simple input functions. These interesting results suggest that we should look next at the response to *periodic* driving functions. For example, what is the stress response of the Maxwell element to a sinusoidal strain? Intuitively we feel that the answer depends on the relative magnitudes of the period of the imposed strain cycle and the relaxation time of the element. In the case of an extremely slow sinusoidal strain, almost total relaxation of the stress occurs at every point on the cycle. The material behaves like a viscous fluid. Any vestigial stress response is almost exactly out of phase with the driving function, since the stress is at its greatest when $d\epsilon/dt$ is at a maximum, that is, when $\epsilon = 0$. If on the other hand the cyclic strain rate is very rapid the dashpot is unable to respond to any extent. The viscous response is entirely prevented and the stress unrelaxed. The amplitude of the induced periodic stress is at a maximum. The material behaves as a purely elastic body. At intermediate frequencies, the stress response is diminished in amplitude and lags in phase. This dynamic viscoelastic behaviour is expressed most succinctly by defining a complex strain

$$\epsilon^* = \epsilon_0 \exp(i\omega t) \tag{3.11}$$

and a complex stress $\sigma^* = \sigma_0 \exp(i\omega t + i\delta)$, where δ is the phase angle between strain and stress. The complex compliance

$$D^* = D' - iD'' = \epsilon^*/\sigma^* = \epsilon_0 \exp(-i\delta)/\sigma_0 \tag{3.12}$$

For a Maxwell element the total rate of strain is simply the sum of the spring and dashpot rates of strain

$$\frac{d\epsilon^*}{dt} = D\frac{d\sigma^*}{dt} + \eta^{-1}\sigma^* \tag{3.13}$$

Here D is the compliance of the spring. By substituting and separating real and imaginary parts we obtain

$$\epsilon_0/\sigma_0 = D\cos\delta + \eta^{-1}\omega^{-1}\sin\delta \tag{3.14}$$

and

$$\tan\delta = 1/\eta\omega D = D''/D' \tag{3.15}$$

By definition (equation 3.12) $D' = \epsilon_0\cos\delta/\sigma_0$ and thus from equations 3.11 and 3.13 it follows that $D' = D$.

Similarly $D'' = \epsilon_0\sin\delta/\sigma_0 = 1/\eta\omega$. σ_0, the amplitude of the stress response at any frequency ω, is given by

$$\sigma_0 = \epsilon_0/(D\cos\delta + \sin\delta \times \eta^{-1}\omega^{-1}) \tag{3.16}$$

At high frequencies $\delta \to 0$ and $\sigma_0 = \epsilon_0/D$. At very low frequencies $\delta \to \pi/2$ and $\sigma_0 = 0$. When $\omega = 1/\eta D$ $\tan\delta = 1$ and $\sigma_0 = \sqrt{2}\epsilon_0/2D$.

Figure 3.9 shows ϵ and σ for the frequencies $\omega = 10^n/\eta D$, $n = -1, 0, 1$.

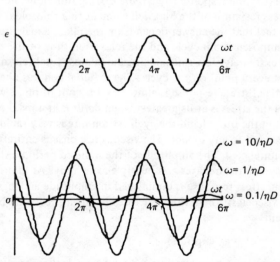

Figure 3.9   Stress response of a Maxwell element to a sinusoidal strain at three different frequencies (*see* equations 3.15 and 3.16). Note the reduction in stress amplitude and the increase in phase angle as frequency falls

The three-element model shown in figure 3.7 has properties resembling those of polymer materials. At high frequencies the parallel combination of spring (compliance D_2) and dashpot behaves as a rigid link, and the system behaves as though it consisted only of the series spring (compliance D_1). At very low frequencies the system reduces to a series combination of springs (compliance $D_1 + D_2$). In general at any frequency ω

$$D' = D_1 + \frac{D_2}{1 + \omega^2 \tau^2}$$

and

$$D'' = \frac{D_2 \omega \tau}{1 + \omega^2 \tau^2}$$

where

$$\tau = D_2 \eta \tag{3.17}$$

When

$$D_2 \eta \omega \gg 1 \quad D' = D_1$$

When

$$D_2 \eta \omega \ll 1 \quad D' = D_1 + D_2$$

When

$$\omega = \tau^{-1} \quad D' = D_1 + D_2/2$$

(figure 3.10). The response of real polymer materials can be accurately modelled only by assuming an extended array of Voigt elements, and a corresponding

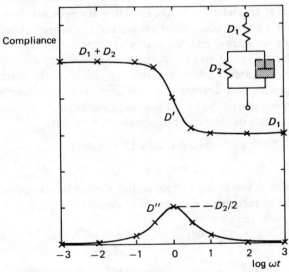

Figure 3.10 Frequency dependence of the storage compliance D' and the loss compliance D'' (components of the complex compliance D^*) calculated for the standard linear solid (arbitrary values $D_1 = 3$ and $D_2 = 2$ are used)

distribution of retardation times τ. This has the effect of broadening the loss peak.

Model analysis (and the generalised compliance D) may be used to represent several kinds of deformation in real solid materials. Shear deformation (involving only changes of shape) and bulk compression (involving only changes of volume) are the fundamental modes. The corresponding compliances are

$$J = \text{Shear strain/Shear stress}$$

$$B = -(\text{Bulk strain/Pressure})$$

Other deformations may be decomposed into combinations of these two. Thus simple extension in response to a uniaxial tensile stress produces both shape and volume changes. The tensile compliance $D = J/3 + B/9$.

Finally we note that the equations developed above may be expressed in terms of modulus rather than compliance. Just as $D = 1/E$ (equations 3.1 and 3.2), so $D^* = 1/E^*$, where E^* is a complex modulus $E^* = E' + iE''$. Once again real solids have distinct shear and compressional moduli, $G^* = 1/J^*$ and $K^* = 1/B^*$, respectively. Most interest in polymer viscoelasticity lies in the mechanics of shear deformation, studied either directly in pure shear or indirectly in extension.

3.12 Energy of Deformation

We turn now to a new consideration. A force acting on a material and producing a deformation performs work. This work is $\int \sigma \, d\epsilon$ per unit volume, where σ is the instantaneous stress and ϵ the strain. The integral represents the energy required to produce deformation. What becomes of this energy of deformation?

In the case of a hookean spring, the energy of deformation is stored indefinitely in the spring and may be completely recovered by allowing the spring to act on another load and return to zero strain. In contrast, the energy of deformation expended on a dashpot is at once converted irrecoverably to heat. The amount of work performed by the load in producing a given displacement in the dashpot depends on the rate of displacement $r = d\epsilon/dt$.

$$\text{Work performed} = \text{Load} \times \text{Displacement}$$

$$= \eta r \times \epsilon$$

The work expended in producing a strain against viscous forces can be made as small as desired by reducing the rate of displacement.

If the load is cyclic then

$$\text{work performed per cycle} = \oint \sigma \frac{d\epsilon}{dt} \, dt$$

$$= \oint \sigma_0 \cos \omega t \, \frac{\sigma_0 \cos \omega t}{\eta} \, dt$$

$$= \pi \sigma_0^2 / \eta \omega \tag{3.18}$$

$$\text{Power required} = \frac{\pi\sigma_0^2}{\eta\omega}\frac{\omega}{2\pi} = \frac{\sigma_0^2}{2\eta} \qquad (3.19)$$

However, as the frequency ω increases, the strain amplitude produced by a given σ_0 decreases. If the spring and dashpot are series coupled to give a Maxwell element, then the energy of deformation is stored at high frequencies and lost at low frequencies. In the three-element model, the energy of deformation is stored at both high and low frequencies and partially lost at intermediate frequencies only. The maximum loss occurs when tan δ and D'' reach their maximal values. It is for this reason that tan δ is often described as the *loss tangent*. The lost energy appears as heat which may produce large increases in temperature in a material subject to continuous cyclic stress at frequencies near to τ^{-1}. In contrast, the energy of *elastic* deformation is zero over a complete cycle.

3.13 Anelastic Mechanical Spectra

A simple device for studying mechanical loss is the torsional pendulum, shown in figure 3.11(a). The decay of the amplitude of the torsional oscillations is exponential: the cyclic strain damped by viscoelastic loss in the polymer rod or thread may be written

$$\epsilon = \epsilon_0 \exp(-\alpha t) \sin (\omega t - \delta) \qquad (3.20)$$

and

$$\tan \delta = 2\alpha/\omega = \Lambda/\pi \qquad (3.21)$$

where Λ is the *logarithmic decrement* $\ln(\epsilon_i/\epsilon_{i+1})$ calculated from the amplitude of successive oscillations — figure 3.11(b).

The frequency of oscillation is determined by the moment of inertia of the disc and may in practice be varied between 0.1 and 10 Hz. It is common to study the dependence of tan δ on temperature at a fixed frequency. Such a plot is known as an anelastic spectrum or loss spectrum — figure 3.11(c). The analysis of the three-element model leads us to expect maxima in tan δ when ω^{-1} is equal to a characteristic relaxation time of the material. Polymer materials invariably show a number of loss peaks, and we identify each one with one of the types of molecular motion described in chapter 2. The relaxation times of each decrease with increasing temperature, so that there is for each at any frequency a maximum in the associated loss process at a characteristic temperature. Alternatively loss spectra can be obtained by studying forced vibrations over a wide range of frequencies at a fixed temperature. Such equipment is of particular value in studying the technical performance of rubbers where the dynamic response to forced vibrations of 100–1000 Hz is of interest.

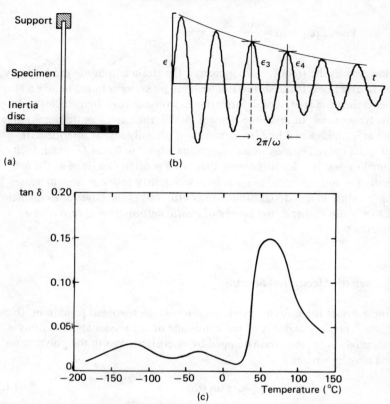

Figure 3.11 (a) Simple torsional pendulum for the study of viscoelastic loss.
(b) Damped sinusoidal strain showing exponential decay of strain amplitude
from which logarithmic decrement and loss tangent may be calculated. (c)
Anelastic spectrum of PA 6

3.14 Time—Temperature Correspondence and the WLF Equation

We consider next how the time (or frequency) dependence of modulus or
compliance variables is related to the temperature dependence. Figure 3.12
shows shear stress relaxation modulus data for an amorphous polymer at two
temperatures T_0 and T_1. The two curves clearly have some similarity of shape
and it is found that they are in fact reducible to a single curve by the application
of a dimensionless scaling factor a_T to the time axis. Thus

$$G(t,T_0) \approx G(t/a_T, T_1) \tag{3.22}$$

a_T, as the subscript indicates, depends on the temperatures T_1 and T_2. This
amounts to saying that stress relaxation modulus curves obtained at different

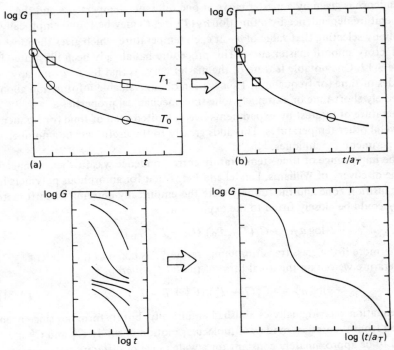

Figure 3.12 Stress relaxation modulus $G(t)$ of an amorphous polymer at two temperatures. The curves have similar shapes and can be made to lie on a common curve by the application of a suitable scaling factor a_T to the time axis.
(b) Formation of a stress relaxation master curve

temperatures lie on a common master curve when plotted against the reduced variable t/a_T. The position of the master curve on the t/a_T axis depends on the choice of an arbitrary reference temperature T_0 for which $a_T = 1$. If a logarithmic time axis is used, the scale factor a_T becomes a horizontal shift $\log a_T$ (*see also* figure 3.12). A minor correction is normally applied to allow for effects arising from changes in density ρ. A reduced relaxation modulus $G^r = T_0\rho_0 G(t, T)/T\rho$ is plotted against t/a_T rather than $G(t, T)$ itself. This method of reduced variables can also be applied to small strain creep compliance data by plotting $JT_0\rho_0/T\rho$ against t/a_T. Dynamic compliance and modulus data are obtained as functions of the frequency ω; in these cases ωa_T is the appropriate reduced variable.

These ideas form the basis of the concept of *time–temperature correspondence*. This has been a very fruitful approach to the analysis of data in polymer viscoelasticity because it enables us to separate the time and temperature dependences of important mechanical variables. The time dependence is given by a master curve at one reference temperature and the temperature dependence by a function $a_T(T)$. $a_T(T)$ may be found empirically, simply by selecting that value of a_T at each temperature which gives the most satisfactory smooth master curve. This procedure has already been illustrated in figure 3.12. One notable feature of the master curve is that it can span many decades of time (or frequency). Thus we can obtain valuable information about extremely short-time or extremely long-time mechanical properties at a temperature of interest by experiments over a limited span of time (or frequency) at several other temperatures. This adds greatly to the usefulness of a number of experimental techniques.

The importance of time—temperature correspondence was further enhanced by the discovery of Williams, Landel and Ferry that for amorphous polymers in the transition zone and the rubbery state the empirically determined shift factors $\log a_T$ could be closely fitted to the expression

$$\log a_T = -C_1(T - T_0)/(C_2 + T - T_0) \tag{3.23}$$

Furthermore if the glass transition temperature T_g is taken as the reference temperature we obtain the usual form of the *WLF equation*

$$\log a_T^g = -C_1^g(T - T_g)/(C_2^g + T - T_g) \tag{3.24}$$

This equation was originally established empirically, but its form has since been derived from theoretical models of molecular motion in solids. C_1^g and C_2^g are found to be approximately constant for a wide range of different polymers. Further discussion of the WLF equation can be found in the texts listed at the end of this chapter.

The molecular basis of time—temperature correspondence may be broadly described as follows. The viscoelastic response of a polymer solid depends on the rates of a variety of molecular motions. The effect of change of temperature is to alter these rates. The implication of time—temperature correspondence is that the various molecular motions controlling the viscoelastic properties *all have the same temperature dependence*. Although not generally true, this assumption appears to be widely valid for amorphous polymers in the transition zone, the rubbery state and the terminal zone.

3.15 Nonlinear Viscoelastic Materials

In earlier sections of this chapter we have made much use of the compliance D, as defined in equation 3.3. The compliance is essentially a parameter of *linear* viscoelasticity theory, as its definition implies that the general strain response

$$\epsilon = f(\sigma, t) \tag{3.25}$$

may be written

$$\epsilon = \sigma D(t) \tag{3.26}$$

which expresses a linear relation between ϵ and σ.

Practical polymers may often be treated as linear viscoelastic materials for relatively small deformations and short times. We have already seen how a creep compliance may be fitted to an empirical linear expression. However, strictly linear behaviour is not generally observed. Figure 3.13 shows creep curves for PP. It can be seen that replotting the data in the form $\epsilon(\sigma, t)/\sigma$ does not yield a single master curve. Under these circumstances, analytical avenues do not lead anywhere very useful, and test results such as creep data must be given in graphical or tabular form. The *isochronous stress–strain curve* is a particularly useful form of presentation which shows the relation between σ and ϵ at some fixed time t. A family of such curves is often displayed on a single graph. The isochronous stress–strain curve of a linear viscoelastic material is itself linear, and so the curvature of lines such as those shown in figure 3.14 is a measure of the nonlinearity of the viscoelastic behaviour of the material. Alternatively the *isometric stress–time curve* may be constructed (figure 3.14).

As an alternative to graphical methods a more elaborate empirical equation may be used, such as that of Findley

$$\epsilon(\sigma, t) = \epsilon_0 \sinh(a\sigma) + bt^n \sinh(c\sigma) \tag{3.27}$$

3.16 Impact Performance

Impact damage is a common kind of failure in plastic components, but impact performance is one of the most difficult of mechanical properties to assess

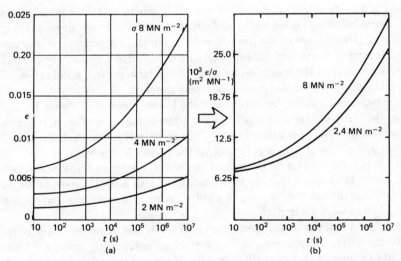

Figure 3.13 (a) Creep curves of PP at 20 °C at three stresses σ. (b) The same data replotted as creep compliance $D = \epsilon(t)/\sigma$ showing that the creep strain does not depend linearly on stress at the two highest values of σ

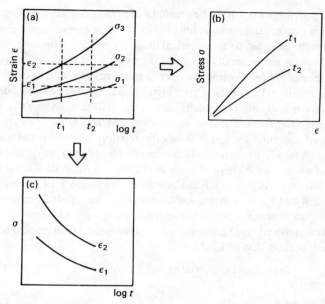

Figure 3.14   Derivation of isochronous stress—strain curves (b) and isometric stress—time curves (c) from a family of creep curves (a)

satisfactorily. It is impracticable to simulate the wide variety of impacts to which an article may be subject in its service life. Many impact test devices have been used, some involving falling balls and darts. The most important impact test machines, however, are of the pendulum type, similar in principle to those used in the testing of metals. Pendulum testers are classified as *Charpy type* or *Izod type*, depending on the way in which the rectangular test piece is held (figure 3.15). In both cases the test piece is normally notched.

The pendulum strikes the test piece at the lowest point of its swing. In failing the test piece takes energy from the down-swinging pendulum, reducing the amplitude of its next upward swing. This amplitude is recorded. The loss in energy of the pendulum is a measure of the *impact strength* (IS) of the material. It is usual to quote the impact strength as *energy to break/fractured area*. It should be noted that the term impact strength is a misnomer, as the word 'strength' usually has the meaning of a critical stress. However, the usage is firmly established in work on impact.

It is found that the impact performance depends greatly on the characteristics of the notch. For example, in tests with blunt notches (tip radius $r = 2$ mm) PVC has a higher impact strength than ABS. If the test pieces are prepared with sharp notches ($r = 0.25$ mm) the order is reversed, and ABS is found to have the higher impact strength — figure 3.16(a). Test results depend also on crack depth

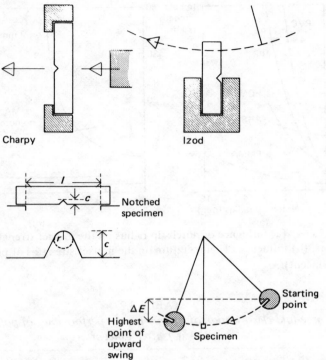

Figure 3.15 Charpy and Izod type pendulum impact tests, showing notching and mounting of the specimen

(c) and for some materials the effects of both c and r on impact strength can be expressed through a simple stress concentration factor $1 + 2(c/r)^{1/2}$ (*see* p. 77). However, the variety of types of impact failure observed, even in a standard Charpy test, is considerable, ranging from brittle failure without any yielding to ductile failure with many intermediate cases. For many design purposes it is satisfactory to rank materials according to a three-fold classification at 20 °C (*see* table 3.2). Impact strength of course may depend strongly on temperature – figure 3.16(b). An arbitrary brittleness temperature T_B is sometimes defined as the temperature at which IS ($r = 0.25$ mm) equals 10 kJ/m^2.

Other factors such as molar mass, presence of additives and processing conditions may influence impact strength. An interesting example is the way that water content affects the impact strength of polyamides (nylons). It is well known that nylons absorb several per cent by weight of water when immersed or conditioned in a humid atmosphere. At room temperature dry nylon is notch brittle with an IS ($r = 0.25$ mm) of about 4 kJ/m^2; wet nylon has an impact strength greater than 20 kJ/m^2 and is classified as tough.

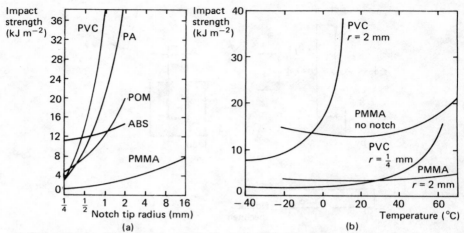

Figure 3.16 (a) Influence of notch tip radius on the impact strength of five polymers. (b) Influence of temperature on the impact strengths of polymers (after Vincent)

TABLE 3.2
Three-fold classification of Charpy impact performance of polymers (20 °C), after Vincent

Brittle	Notch brittle	Tough
Acrylics	PP	Wet nylon
Polymethylpentene	Cellulosics	LDPE
Glass-filled nylon	PVC	ABS (some)
PS and HIPS	Dry nylon	PC (some)
	Acetals	PTFE
	Polysulphone	Some high
	Toughened acrylics	impact PPs
	HDPE	
	PPO	
	ABS (some)	
	PC (some)	
	PETP	

Notes:
Brittle − specimens break even when unnotched.
Notch brittle − specimens do not break unnotched, but break when sharp-notched
          (r = 0.25 mm).
Tough − specimens do not break completely even when sharp-notched (r = 0.25 mm).

3.17 Yield and Fracture

By the criteria of the previous paragraphs, a tough polymer is one which has a high energy to break in an impact test and which fails in a ductile manner. A traditional metallurgical approach to toughness is through the stress—strain relation. The toughness of a metal is measured by the area beneath its stress—strain curve taken to failure. Toughness is associated with ductility, and both the continuum mechanics and the micromechanics of plasticity in metals are now well understood. The search for improved impact performance has more recently stimulated a similar systematic study of yield and fracture processes in polymers. Yield phenomena are exploited in polymer technology in such operations as fibre-drawing and cold-forming (*see* chapter 6), adding further to the significance of basic research in this subject.

In general a polymer may show either brittle or ductile behaviour in a mechanical test according to circumstances. The main controlling factors are the time scale and the configuration of the test procedure, and the temperature. Low rates of straining and high temperatures favour a ductile response — figure 3.17(a, b, c). The Charpy impact method is effectively a high speed test of

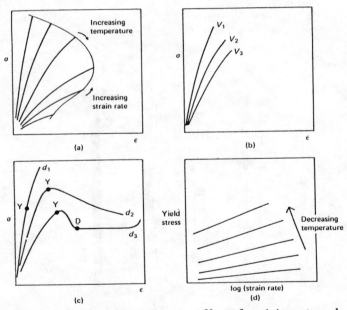

Figure 3.17   (a) Failure envelope showing effect of straining rate and temperature on the stress—strain curve of polymers. (b) Influence of straining rate on tensile strength of a brittle polymer ($V_1 > V_2 > V_3$). (c) Types of ductile behaviour in polymers. (d) Influence of testing speed on the tensile yield stress of an amorphous polymer at several temperatures.

failure in flexure. At the other extreme of time scale are high load creep tests continued to failure. The conventional tensile test may be carried out over a wide range of fixed straining rates. We have already described the characteristics of brittle failure under tensile stress in section 3.1; the material fails at relatively low strain with little or no yield. The *tensile strength* is the nominal stress at failure; for polymers this is not a single-valued property of the material at each temperature since it depends on the rate of straining.

In the case of ductile behaviour the stress–strain curve takes one of the several forms represented by d_1, d_2 and d_3 in figure 3.17(c). In d_1, a decrease in slope at Y marks a more or less well-defined yield point which is followed by uniform yielding until failure. In d_2 a maximum in the nominal stress develops at Y leading to the formation of an unstable neck. As the specimen is further extended it continues to yield inhomogeneously at the neck until failure occurs. In d_3 the formation of a neck at Y is followed by strain-hardening which stabilises the neck. This is the type of inhomogeneous yield behaviour which gives rise to drawing. Extension of the specimen beyond D occurs by increasing the length of the necked portion at constant cross-section (figure 3.18). This proceeds until a limiting overall strain or *natural draw ratio* is attained, after which the fully drawn polymer moves up the final part of the curve to failure.

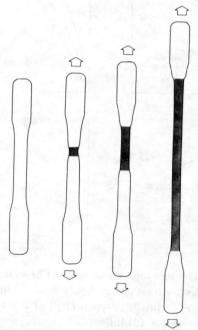

Figure 3.18   Ductile behaviour in a cold-drawing polymer

At temperatures far below T_g polymers show brittle behaviour. In any particular test a ductile–brittle transition may be observed at a temperature T_b (not the same as the impact brittleness temperature T_B). T_b is well below T_g. Between T_b and T_g the yield stress σ_Y falls with rising temperature and approaches zero in the vicinity of T_g itself.

Studies of yield stress in metals have confirmed the general validity of the von Mises yield criterion, according to which yielding occurs when the total shear strain energy reaches a critical value. In metals yield stress is approximately independent of hydrostatic pressure which produces dilatational strain only. In polymers the yield stress increases with hydrostatic pressure. The von Mises yield criterion therefore requires modification for polymeric materials. Both shear and hydrostatic stresses exist in uniaxial tension and compression and in flexure so that observed yield stresses are found to be dependent on the test configuration. One important method of presenting long-term strength data is by means of high load creep curves or the creep failure stress curves derived from such creep tests (figure 3.19).

In this brief discussion of yielding we have so far considered only the macroscopic or continuum features of plasticity. However, the occurrence of ductile behaviour in glassy polymers below T_g must be considered somewhat unexpected from a molecular point of view. How do glassy polymers deform to large strains without fracture at temperatures at which large-scale molecular motion is normally non-existent? This question is not yet finally resolved. The influence of hydrostatic pressure on yield stress once again directs attention to the free volume; at high tensile stresses the hydrostatic tension component may cause the free volume to increase and permit segmental motion in the solid. T_g itself is known to rise with increasing hydrostatic pressure. The stress field itself may assist viscous flow processes by favouring conformations which extend chains in the direction of the applied stress. A model of the morphological changes accompanying fibre-drawing in crystalline polymers has already been described in chapter 2.

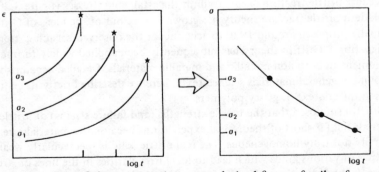

Figure 3.19   Creep failure stress–time curve derived from a family of creep failure curves. * marks the time of failure (rupture) on the creep curves

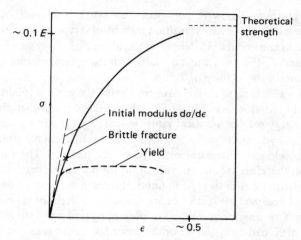

Figure 3.20   Limitation of strength of materials imposed by yield and fracture processes. Simple models of ideal solids predict theoretical strengths of ∼0.1E and failure strains of 0.2–0.5

Yield and fracture are of great practical interest for two reasons. For engineers it is the occurrence of these processes which set limit states for design: critical stresses and strains which must not be exceeded in performance. To materials scientists yield and fracture are processes which generally prevent materials attaining anything approaching their theoretical strengths. As we have already pointed out in figure 3.2 the small strain elastic modulus of a solid material is determined by the interatomic force field. Real materials generally show initial moduli not very different from the theoretical predictions. However, these same materials usually fail to attain the theoretical strengths calculated from the bond energies either because plastic deformation intervenes prematurely or because brittle fracture occurs to limit the attainable stresses (figure 3.20).

Modern brittle fracture theory has grown largely out of the ideas of Griffith, originally proposed around 1920 to account for the observed breaking strengths of glass fibres. Griffith theory has subsequently been applied widely to fracture phenomena in both non-metallic and metallic materials. Detailed accounts will be found in general materials science texts: here we describe briefly its central ideas in the context of glassy polymers.

Griffith supposed that the tensile strengths (and failure strains) of brittle materials fell far short of theoretical expectations because the materials are generally not truly homogeneous. The real brittle solid is spectacularly weakened by *flaws* of various kinds which lead to local irregularities in the lines of stress (figure 3.21). If the flaw takes the form of a crack transverse to the direction of the stress σ, stress concentrations σ_c develop at the ends a, a' of the crack.

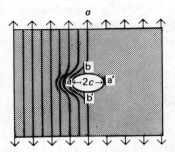

Figure 3.21 Stress concentration at the ends *a* and *a'* of an elliptical crack in a uniformly stressed plate. Stress is reduced in the vicinity of *b* and *b'*; the net reduction in strain energy in a plate of unit thickness due to the presence of the flaw $\Delta_1 = \pi c^2 \sigma^2 / E$. The flaw extends spontaneously if $d\Delta_1/dc \geqslant d\Delta_2/dc$, where Δ_2 is the surface work of fracture (*see* text)

Griffith made use of the equations previously established by Inglis for the stress concentration about an elliptical hole in a uniformly loaded plate. The stress concentration factor $2(c/r)^{\frac{1}{2}}$ associated with a long microcrack of very small radius r at the tip can clearly be very large. If σ_c is sufficiently great to cause local failure then the crack propagates and brittle fracture ensues. Next we ask: what determines the critical value of σ_c? Griffith expressed the criterion of crack propagation in terms of energy and argued that the crack advances only if the strain energy released in extending the crack equals or exceeds the energy required to form the new fracture surfaces. The presence of the crack reduces the total strain energy by an amount Δ_1. Δ_1 may be calculated from the distribution of stress about the crack (figure 3.21): thus

$$\Delta_1 = \pi c^2 \sigma^2 / E \tag{3.28}$$

If Δ_2 is controlled by the specific surface energy γ of the solid

$$\Delta_2 = 4c\gamma \tag{3.29}$$

Therefore

$$\sigma_f = (2E\gamma/\pi c)^{\frac{1}{2}} \tag{3.30}$$

is the critical fracture stress. From breaking strength data it appears that, for glass, Δ_2 is indeed largely determined by γ. For metals and glassy polymers, on the other hand, the formation of new fracture surface involves processes of plastic deformation in the region of the advancing crack tip. Thus the quantity Δ_2 represents a generalised surface work term, which embraces all those processes involved in crack extension. Likewise γ is interpreted as a generalised *fracture surface energy*.

What of the cracks or flaws themselves? In general, brittle fracture may arise from stress concentrations either associated with holes or notches deliberately formed in the material or associated with natural flaws present in the microstructure. In glass, the natural flaws which initiate fracture are usually surface scratches. If these are absent very high tensile strengths can be achieved. Similar exceptional strengths can be attained in ceramic whiskers and fibres of high structural perfection. In metals, the flaws naturally present may arise from the movement of dislocations. What is the nature of the flaws in glassy polymers? Under many circumstances a deliberately introduced crack in a glassy polymer propagates by the formation of a *craze* which runs ahead of the crack tip. The craze material comprises fibrils and microvoids in approximately equal volumes. Crazing is a characteristic form of microyielding associated with brittle fracture. Craze lines are easily observed by the scattering of light from refractive index discontinuities produced by the microvoids. Brittle fracture in materials free of deliberate cracks may proceed by the formation at one or more microscopic inhomogeneities (including voids and inclusions) of craze material, followed by rupture, followed by propagation of the initiated crack. Craze lines are formed at right angles to the stress in glassy polymers such as PS under high uniaxial tensile stress as a prelude to brittle fracture.

A large body of experimental and theoretical work now exists on the fracture mechanics of polymers. The generalised $E\gamma$ of equation 3.30 is designated K, the stress intensity factor. The critical value of K, K_c, is therefore a measure of the fracture toughness of the material, an intrinsic material property. The quantity $G_c = K_c^2/E$, the strain energy release rate (per unit area of fracture, not per unit time) is an essentially equivalent measure of material fracture toughness. These quantities are usually measured in controlled crack growth experiments. Data for some polymers are collected in table 3.3. The traditional Izod and Charpy impact tests (*see* figure 3.15 and text) have been analysed in terms of linear fracture mechanics theory, with the result that it is now possible to derive K_c (or G_c) data from such tests, provided that the notch depth can be varied. The energy absorbed on impact $\Delta E = G_c \times bd\phi$, where b is the width of the sample, d the thickness and ϕ is a geometrical factor depending on c/d.

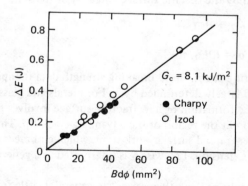

Figure 3.22 Charpy and Izod impact test data for medium density polyethylene at 20 °C (after Williams)

TABLE 3.3

Izod fracture toughness G_c *(kJ/m)*
(after Williams)

Polystyrene	0.8
Poly(methyl methacrylate)	1.4
Poly(vinyl chloride)	1.4
HDPE	3.1
Polycarbonate	4.8
Polyamide PA-66	5.0
PE (rel density 0.94)	8.4
High impact polystyrene	14
LDPE	34
ABS	47

Plotting ΔE against $bd\phi$ for various values of c allows G_c to be calculated (figure 3.22).

3.18 Friction of Polymers

Friction and wear characteristics are important but complicated aspects of the mechanics of polymers. Knowledge of both is fundamental to the design of rubber vehicle tyres, but the general significance of friction and wear has widened as many other polymers have found engineering application as moving parts. Frictional forces developed between the moving fibres in textile manufacturing processes have also been much studied. The whole subject is a complex one, lacking firmly established quantitative theories.

The classical engineering laws of sliding friction are simple. Amontons' laws state that the friction F between a body and a plane surface is proportional to the total load L and independent of the area of contact A. The friction of a moving body is normally somewhat less than that of a static body. The kinetic friction is considered to be independent of velocity. The *coefficient of friction* μ is defined as the ratio F/L. Both materials (slider and surface) must be specified. Coefficients of friction for a number of materials are given in table 3.3. Polymers as a class exhibit a wide range of friction properties. PTFE is particularly used for its remarkably low friction, whereas rubbers possess high values of μ.

Amontons' laws hold only approximately for metals and frequently do not hold for polymers. The facts can be explained only in terms of microscopic surface features of the solids. First we accept that the surfaces of materials are not atomically flat but microscopically irregular. Therefore the true area of contact between two surfaces is very much less than the apparent area of contact, and the entire normal force is carried by the tips of the asperities. At these contact points the local stress is so great that severe deformation occurs, and the

TABLE 3.4
Coefficients of friction

	μ_k coefficient of sliding friction
PTFE	0.04–0.15
LDPE	0.30–0.80
HDPE	0.08–0.20
PP	0.67
PS	0.33–0.5
PMMA	0.25–0.50
PETP	0.20–0.30
PA 66	0.15–0.40
PVC	0.20–0.90
PVDC	0.68–1.80
PVF	0.10–0.30
SBR	0.5–3.0
BR	0.4–1.5
NR	0.5–3.0

Typical reported values; sliding on various
countersurfaces

tip of each asperity is crushed to produce a small plane or almost plane region. Over this small region intimate atomic contact exists between the two surfaces and *adhesion* occurs (van der Waals' forces or stronger chemical forces act across the contact faces). In order to move the body across the surface it is now necessary either to break the adhesional bonds at the interface AA′ (figure 3.23) or to shear one or other of the materials at some plane (BB′ or CC′) very near the surface. These ideas lead to a basic concept of adhesional friction: that the friction $F = As$, where A is the true area of contact and s is the shear strength of the material. To the extent that soft materials are weak and hard materials strong, the effects of factor A and s are in opposition (figure 3.23). This is why such diverse engineering materials as metals, ceramics and polymers may have surprisingly similar coefficients of friction.

In the case of metals the true area of contact produced by *plastic* deformation of surface asperities is roughly proportional to the load, so that the first law holds. Generally this law does not hold for polymers. First, for rubbers the deformation is largely elastic rather than plastic and the area of contact increases with $L^{2/3}$ rather than L. Thus the coefficient of friction decreases with

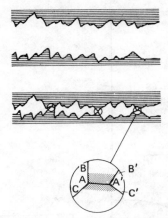

Figure 3.23 Micromechanics of friction: adhesional contact occurs at the asperities and sliding motion requires shearing at or near the interface

pressure. Second, for polymers which deform viscoelastically the friction obeys a law of the form

$$F = kL^x$$

(3.31)

or

$$\mu = kL^{x-1}$$

with $\frac{2}{3} < x < 1$. For example x has the value 0.85 for PTFE over 10 orders of magnitude in L.

The shearing processes dissipate energy to an extent that depends on the viscoelastic properties of the material, and hence depends on both the temperature and the strain rate. Figure 3.24 shows that the coefficient of friction of crystalline polymers above T_g depends strongly on the sliding speed, and passes through a maximum. These curves are shifted to the right as the temperature rises. Grosch showed that similar effects are observed in rubbers sliding on smooth glass surfaces, and he demonstrated the viscoelastic origin of the friction by showing that the WLF shift (*see* section 3.14) condenses the data into a single friction–temperature–velocity master curve (figure 3.24). Furthermore, the positions of the peak in the friction master curve and the peak in the loss tangent master curve for each rubber were such that the ratio $V/2\pi\omega$ is about 6 nm, pointing directly to the molecular scale of the energy absorption process.

If two materials of very different hardness are sliding against each other the adhesional model of friction is incomplete. This situation is common in polymer applications where polymer materials bear on metal surfaces. The asperities of the harder material plough into the surface of the softer, producing grooves which may recover as the ploughing tip moves on or from which material may be

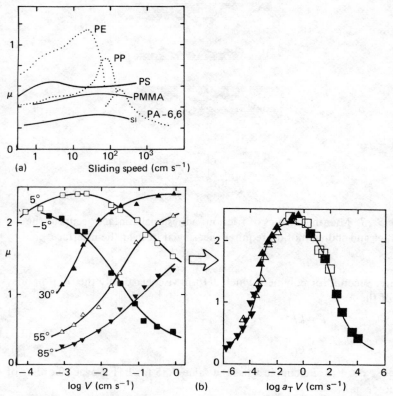

Figure 3.24 (a) The coefficient of friction μ of six polymers, showing dependence on sliding speed at room temperature (after McLaren and Tabor). (b) Coefficient of friction of acrylonitrile–butadiene elastomer on wavy glass as a function of sliding speed V at several temperatures; and master curve obtained by application of WLF shift (after Grosch)

torn out completely. In either case energy is absorbed by the deformed surface and partially dissipated through viscoelastic loss. There are both adhesional and deformational contributions to the energy of sliding one surface across another, and hence to the friction. We thus have two processes absorbing energy and producing friction, and both are expected to be sensitive to strain rate and temperature.

PTFE has an exceptionally low coefficient of friction ($\mu = 0.06$). This material is apparently able to form only weak adhesional bonds (cf. section 6.7), but this fact alone seems an inadequate explanation. HDPE has a coefficient of friction ($\mu = 0.08$) almost as low as that of PTFE, whereas LDPE shows a value of ~0.3. Furthermore, for both PTFE and HDPE μ rises to ~0.3 at higher sliding speeds. It is probable that some unusual mode of shearing occurs at the

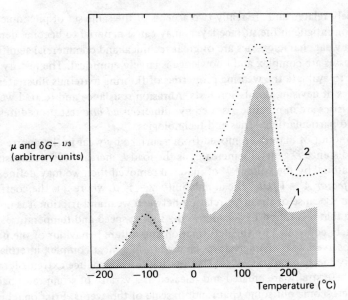

Figure 3.25 Temperature dependence of rolling friction and of hysteresis loss of PTFE compared. Curve 1 (shaded): the quantity $\delta G^{-1/3}$ measures the mechanical energy loss per cycle (δ is the loss angle and G the torsional modulus) Curve 2 (dotted): μ the coefficient of friction (After Ludema and Tabor)

polymer surface during sliding, and that this can occur because of the regular and unbranched molecular structure.

Rolling friction is determined largely by deformational energy losses since the tensile strength (rather than the shear strength) of the adhesion in the contact regions is low. A correlation is usually found between the coefficient of rolling friction and other viscoelastic properties (*see* figure 3.25). Rolling friction has much in common with lubricated sliding for in this case the thin liquid film of lubricant between the surfaces provides a plane of extremely low shear strength, and once again deformational processes dominate the friction. The observed coefficient of friction of rubbers depends on the loss: high loss rubbers show higher coefficients of friction than low loss rubbers of the same hardness in both rolling and lubricated sliding. In consequence high loss rubbers have been used to improve the skid resistance of motor vehicle tyres.

3.19 Wear

Both the shearing of adhesive bonds and the ploughing of deformations may remove material from the surface of the softer material. Wear and friction are thus closely related and are really two aspects of the same set of phenomena. In addition fatigue in the surface layers may cause material to become detached. Although wear characteristics are of great technical and commercial significance the processes are complex and knowledge is largely empirical. The history of attempts to evaluate the wearing properties of flooring materials illustrates the difficulties of devising simulation tests. Abrasion resistance and related wear characteristics are in practice assessed by numerous *ad hoc* test procedures, specific to particular industries and technologies.

The amount of material removed from a surface by wear in the course of its sliding a distance D generally increases as the load L increases. If we measure the loss of material by the volume V of material removed then we may define the *abrasion factor $A' = V/DL$*. The abradability $\gamma = A'/\mu$, where μ is the coefficient of friction. Because of the close relation between wear and friction it is not surprising that γ is found to depend on the sliding speed and temperature. Wear is ultimately controlled by the deformation and failure behaviour of the polymer.

The observed wear characteristics emerge from a most complex interplay of elementary mechanical processes. The wearing forces produce extreme local strains. These strains are created and released at a variety of strain rates because of the microscopic but somewhat random scale of the events. Frictional heating introduces large local changes in temperature, which may markedly alter the viscoelastic response of the material. Finally, the phenomena are controlled by the nature of the *surfaces* of the materials, which frequently differ from the bulk solid.

Despite these difficulties the ideas summarised here give considerable insight into the nature of wear and abrasion. They are of great value in unifying the results of diverse technical tests, suggesting correlations between experimental parameters and rationalising test results. Wear, like metallic corrosion and the degradation of polymers by weathering processes, is immensely wasteful of materials and human effort. Its study forms part of the science of tribology (and more generally of terotechnology).

3.20 Heat Transmission

Innumerable engineering uses of polymers involve the transmission of heat. In some cases this is an incidental effect, as when rubber vehicle tyres become hot by viscoelastic loss. In other situations the control of heat flow is a primary consideration. For example polymer materials are used as thermal insulants in many industries. In every case the design engineer must consider the working temperatures reached within these materials. We have seen earlier how sensitive are the mechanical properties of polymers to changes of temperature. We shall see in chapter 5 that excessive heating can cause serious irreversible deterioration of polymers through chemical degradation processes. The thermal properties discussed here also have an important bearing on the fire behaviour of polymers.

The methods of analysis of heat flow are much the same whatever the nature of the material and depend on a knowledge of two material properties, the thermal conductivity λ and the specific heat capacity (or specific heat) c_p.

The heat flow q at any point in a solid is proportional to the temperature gradient

$$q = -\lambda \, \frac{d\theta}{dx}$$

Thus if the surfaces of a slab are maintained at temperatures θ_1 and θ_2 then the *steady* heat flow is

$$q = \lambda a(\theta_1 - \theta_2)/d$$

where a is the slab area and d the thickness. For a series of similar slabs of different materials the rates at which heat is transmitted are proportional to the thermal conductivities.

If we consider transient rather than steady flows then rates of change of temperature within the solid are determined by the quantity

$$\alpha = \lambda/c_p \rho$$

rather than λ alone. α is the *thermal diffusivity*. How quickly the temperature rises when heat flows into a material at a particular rate depends inversely on ρc_p, the heat capacity of unit volume. We shall discuss first λ and then briefly c_p.

3.21 Thermal Conductivity

At atomic level the effect of applying heat at one face of a cold slab is to increase the amplitude of thermal vibration in that face. This additional thermal energy then diffuses in the direction of the opposite face at a rate which for non-metals depends on how strongly the vibrational motion of adjacent atoms and groups is coupled. Strong coupling occurs in covalently bonded materials and transmission of heat is relatively efficient in ordered crystalline lattices. Thus, materials such as crystalline silica (quartz) and diamond, in which all atoms are built into the crystal structure by strong covalencies, are good thermal conductors. This is particularly so at low temperatures where such materials compare with metals in thermal conductivity. As the temperature rises, the thermal motion of the lattice offers increasing resistance to the flow of heat and λ falls. Resistance to the flow of heat also arises from defects in the crystal structure. An extreme of disorder is reached in amorphous solids, which show characteristically low thermal conductivities (compare crystalline and amorphous silica in figure 3.26). Molecular solids in which secondary chemical forces bind the crystal structure together conduct heat poorly because of the weak coupling between molecules. Elementary solid state theory shows that $\lambda = c_p(\rho K)^{1/2} l$ where

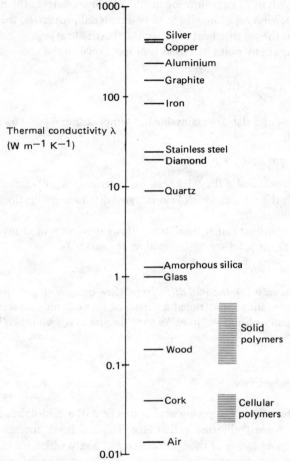

Figure 3.26   Thermal conductivity λ of selected materials

K is the elastic bulk modulus and l the mean free path of the thermal vibrations (phonons). For polymers taking $K \approx 10^3$ MN m^{-2} and $l \approx 200$ pm (the separation of adjacent polymer chains) we obtain $\lambda \approx 0.3$ W m^{-1} K^{-1}, in broad agreement with observed values.

In the case of metals at normal working temperatures the contribution of atomic lattice motion to the thermal conductivity is masked by the much more efficient transmission of thermal energy from one point to another by the mobile electrons. Hence the well-known Wiedemann–Franz law that thermal and electrical conductivities of metals are proportional. Except at very low temperatures metals exhibit much higher thermal conductivities than dielectric solids. Metallic thermal conductivity decreases with temperature because electrons are scattered by lattice vibrations. λ is also reduced by defects, such as are introduced by certain types of alloying. Figure 3.26 shows thermal conductivities of a variety of engineering materials, illustrating these general ideas.

Table 3.5 lists data for individual polymers. What emerges from a detailed study of such data may be summarised as follows.

(1) The range in λ found in solid polymers is small, all values lying within a factor of 2 of 0.22 W m^{-1} K^{-1}.

TABLE 3.5
Thermal properties of polymer materials

	Thermal linear expansivity (10^{-5} K^{-1})	Specific heat capacity (kJ kg^{-1} K^{-1})	Thermal conductivity (W m^{-1} K^{-1})
PMMA	4.5	1.39	0.19
PS	6–8	1.20	0.16
PUR	10–20	1.76	0.31
PVC unplasticised	5–18.5	1.05	0.16
PVC 35% plasticiser	7–25	–	0.15
LDPE	13–20	1.90	0.35
HDPE	11–13	2.31	0.44
PP	6–10	1.93	0.24
POM copolymer	10	1.47	0.23
PA 6	6	1.60	0.31
PA 66	9	1.70	0.25
PETP	–	1.01	0.14
PTFE	10	1.05	0.27
PCTFE	5	0.92	0.14
EP	6	1.05	0.17
CR	24	1.70	0.21
NR	–	1.92	0.18
Fluorocarbon elastomer	16	1.66	0.23
Polyester elastomer	17–21	–	–
PIB	–	1.95	–
Polyethersulphone	5.5	1.12	0.18

Note: A comparison of published data on thermal properties shows numerous inconsistencies and uncertainties. The table contains selected values which should be regarded as indicative rather than firm.

(2) Crystalline polymers (PE, PP, PTFE, POM) show somewhat higher λ than amorphous polymers (PVC, PMMA, PS, EP).

(3) For most crystalline polymers, λ increases with density and crystallinity (for example, compare LDPE and HDPE).

(4) For amorphous polymers, λ increases with molar mass (that is, with chain length) because thermal energy flows more freely along polymer chains than between them. Similarly, addition of low molar mass plasticiser reduces λ for PVC.

(5) λ increases with temperature for some polymers, decreases for others. λ rarely changes by more than 10 per cent from 0 to 100 °C.

(6) The alignment of polymer chains by stretching produces an anisotropy in λ. λ for axial heat flow increases whereas for transverse flow it decreases. In the case of PVC at 300 per cent elongation, heat flows almost twice as rapidly along the axis as across it. High density PE exhibits a tenfold increase in λ along the axis of orientation at 1000 per cent strain. Similar effects have been observed in some rubbers.

Extremely low thermal conductivities can be achieved in cellular polymers; some representative data are given in table 3.6. For high and medium density foams, the thermal conductivity decreases as the density falls. λ is roughly the weighted mean of the thermal conductivity of solid polymer and blowing gas. λ of the foam may be significantly reduced by using a halocarbon gas

TABLE 3.6

Thermal conductivity λ of cellular polymer materials

	Density $(kg\ m^{-3})$	λ $(W\ m^{-1}\ K^{-1})$
PS	16	0.039
	25	0.035
	32	0.032
PVC	35	0.028
	45	0.035
UF	8	0.030
PUR	16	0.040
	32	0.023
	64	0.025
	96	0.043
PE	38	0.046

($\lambda = 0.009$ W m^{-1} K^{-1}) rather than air ($\lambda = 0.024$ W m^{-1} K^{-1}) for foaming. Such materials normally age as air diffuses slowly into the cells over weeks or months, causing the thermal conductivity to rise. In low density foams, heat transfer by convection and radiation becomes significant, and at the lowest densities (< 30 kg m^{-3}) the thermal conductivity once again rises.

Convection is only important in cells larger than about 5 mm in diameter. Radiative heat transfer was found to account for about 20 per cent of the total heat flow in a cellular PS at 30 °C. Both scattering and absorption (*see* chapter 5) of infrared radiation occur within the foam, and the rate of radiative heat transfer depends in a complicated way on cell size and polymer composition. Its importance increases rapidly as the temperature rises since the efficiency of radiative transfer increases approximately as T^2.

3.22 Specific Heat Capacity

Polymer data are given in table 3.5, and data for other engineering materials in table 3.7. c_p is determined largely by the chemical structure rather than the microstructure, and does not range widely among polymers. The amount of heat

TABLE 3.7
Specific heat capacity and expansivity of selected materials

	Thermal linear expansivity α (K^{-1} x 10^{-5})	Specific heat capacity c_p (kJ kg^{-1} K^{-1})
Brass	2.0	0.38
Carbon steel	1.1	0.48
Aluminium	2.3	0.88
Copper	1.7	0.38
Lead	2.9	0.13
Borosilicate glass	0.3	0.78
Building brick	0.5	0.8
Alumina	0.8	
Water	7 (liquid) 5 (ice)	4.2
Solid polymer materials	4–20	1–2

required to produce a particular rise in temperature depends on the vibrational and rotational motions excited within the solid. Replacement of hydrogen by heavier atoms such as F or Cl causes c_p to fall. Between the glass transition and melting temperature recrystallisation may occur, releasing heat from within the solid and making heating rates somewhat irregular.

3.23 Thermal Expansivity

Polymers as a class show higher expansivities than most metals and ceramics. The expansion of polymers is not usually a truly linear function of temperature (α is not constant). For example, the expansivity of PS increases rather erratically by about 50 per cent between 0 and 100 °C. The reason for differences in α between polymers has been little studied. However, engineering data are readily available. Changes in composition can produce significant changes in α. For example, the expansivity of epoxy resins is generally about 5×10^{-5} K^{-1}. Inorganic and metallic fillers can of course be incorporated to reduce α. The flexible formulations may show values as high as 20×10^{-5} K^{-1}. In some applications in which large temperature changes must be tolerated the designer may need to seek a compromise between modulus and expansivity to minimise thermal stresses.

3.24 Melting Range, Glass Transition Temperature and Softening Point

Semicrystalline polymers do not exhibit sharp melting points but fairly well defined melting ranges which may be measured by the methods of light microscopy, X-ray diffraction and scanning calorimetry. For both amorphous and crystalline polymers glass transition temperatures may be determined from changes in a variety of physical properties (*see* section 2.1). There exist also several simple standard test procedures to determine single-point properties of polymers such as the *heat distortion* or *deflection temperature* or the *softening point*. These are directly related to mechanical behaviour and have some limited comparative value.

3.25 Melt Properties

The mechanical properties of the polymer melt determine how easily a polymer may be melt processed (*see* chapter 6). High melt viscosities make extrusion and injection moulding difficult. The melt viscosity, measured by a rotating cylinder, rotating cone or capillary viscometer, is therefore a key indicator of 'processability': for polyolefins, the *melt flow index* is the standard measure, calculated from the mass of polymer which flows through a test orifice under specified load at a prescribed temperature.

While it is useful to talk of a simple melt viscosity, it would be very misleading to imply that molten polymers are rheologically simple. Detailed analysis of polymer melt flow (such as arises in the analysis of melt processing operations and tool design) has to take full account of their complex elastico-viscous properties. Just as polymer solids below T_m show creep and stress relaxation, so polymer melts show some elastic character. An important practical manifestation of this appears in the extrusion process. The reduction in compressive stress acting on the extruded polymer as it emerges from the die produces an increase in transverse dimensions: so-called *extrudate swell* or *die swell*. It is desirable to allow accurately for this elastic recovery in the design of extrusion dies.

Suggestions for Reading

Mechanical Properties

Aklonis, J. J. and MacKnight, W. J., *Introduction to Polymer Viscoelasticity*, 2nd edn (Wiley—Interscience, New York, 1983).

American Society for Metals, *Polymeric Materials: Relationships between Structure and Mechanical Behaviour*, chs 4, 7 and 9 (ASM, Metals Park, Ohio, 1975).

Andrews, E. H., *Fracture in Polymers* (Oliver & Boyd, Edinburgh, 1968).

Andrews, E. H., 'Fracture', in A. D. Jenkins (Ed.), *Polymer Science: A Materials Science Handbook*, pp. 579–620 (North-Holland, Amsterdam, 1972).

Arridge, R. G. C., *An Introduction to Polymer Mechanics* (Taylor and Francis, London, 1985).

Bartenev, G. M. and Lavrentev, V. V., *Friction and Wear of Polymers* (Elsevier, Amsterdam, 1981).

Berry, J. P., 'Fracture of polymeric glasses', in H. Liebowitz (Ed.), *Fracture: An Advanced Treatise*, vol. 7, pp. 37-92 (Academic Press, New York, 1972).

Bowden, F. P. and Tabor, D., *Friction and Lubrication* (Methuen, London, 1967).

Briscoe, B. J. and Tabor, D., 'Friction and wear of polymers', in D. T. Clark and W. J. Feast (Eds), *Polymer Surfaces*, ch. 1 (Wiley, Chichester, 1978).

Brown, R. P., *Physical Testing of Rubbers* (Applied Science, London, 1979).

Brown, R. P. (Ed.), *Handbook of Plastics Test Methods*, 2nd edn (Godwin, London, 1981).

Bucknall, C. B., Gotham, K. V. and Vincent, P. I., 'Fracture — the empirical approach', in A. D. Jenkins (Ed.), *Polymer Science: A Materials Science Handbook*, pp. 621-685 (North-Holland, Amsterdam, 1972).

Cherry, B. W., *Polymer Surfaces* (Cambridge University Press, 1981).

Ferry, J. D., *Viscoelastic Properties of Polymers*, 2nd edn (Wiley, New York, 1970).

Hertzberg, R. W. and Manson, J. A., 'Fracture and fatigue', in *Encyclopaedia of Polymer Science and Engineering*, 2nd edn, vol. 8, pp. 328–453 (Wiley, New York, 1987).

Kambour, R. P., 'Crazing', in *Encyclopaedia of Polymer Science and Engineering*, vol. 4, pp. 299–323 (Wiley, New York, 1986).

Kambour, R. P. and Robertson, R. E., 'The mechanical properties of plastics', in A. D. Jenkins (Ed.), *Polymer Science: A Materials Science Handbook*, pp. 687–822 (North-Holland, Amsterdam, 1972).

Kinloch, A. J. and Young, R. J., *Fracture Behaviour of Polymers* (Applied Science, London, 1983).

Lancaster, J. K., 'Abrasion and wear', in *Encyclopaedia of Polymer Science and Engineering*, 2nd edn, vol. 1, pp. 1–35 (Wiley, New York, 1985).

Lee, L.-H. (Ed.), *Polymer Wear and its Control* (American Chemical Society, Washington DC, 1985).

Moore, D. F., *The Friction and Lubrication of Elastomers* (Pergamon, Oxford, 1972).

Moore, D. F., *Principles and Applications of Tribology* (Pergamon, Oxford, 1975).

Moore, D. F., *The Friction of Pneumatic Tyres* (Elsevier, Amsterdam, 1975).

Nielsen, L. E., *Mechanical Properties of Polymers and Composites*, 2 vols, (Dekker, New York, 1974).

Ogorkiewicz, R. M. (Ed.), *Engineering Properties of Thermoplastics* (Wiley-Interscience, London, 1970).

Pomeroy, C. D., (Ed.), *Creep of Engineering Materials* (Mechanical Engineering Publications, London, 1978).

Powell, P. C., *Engineering with Polymers* (Chapman and Hall, London, 1983).

Schultz, J. M., 'Fatigue behaviour of engineering polymers', *Treatise on Materials Science and Technology*, vol. 10 pt B, pp. 599–636 (Academic Press, New York, 1977).

Sternstein, S., 'Mechanical properties of glassy polymers' in J. M. Schultz (Ed.), *Treatise on Materials Science and Technology*, vol. 10, pt B, pp. 541–598 (Academic Press, New York, 1977).

Treloar, L. R. G., *The Physics of Rubber Elasticity*, 3rd edn (Oxford University Press, 1975).

Turner, S., *Mechanical Testing of Plastics*, 2nd edn (Godwin, London, 1984).

Vincent, P. I., *Impact Tests and Service Performance of Thermoplastics* (Plastics Institute, London, 1971).

Ward, I. M., *Mechanical Properties of Solid Polymers*, 2nd edn (Wiley-Interscience, London, 1983).

Ward, I. M., 'The preparation, structure and properties of ultra-high modulus flexible polymers', in *Advances in Polymer Science*, vol. 70, *Key Polymers – Properties and Performance*, pp. 1–70 (Springer, Berlin, 1985).

Ward, I. M., 'High modulus flexible polymers', in L. A. Kleintjens and P. J. Lemstra (Eds), *Integration of Fundamental Polymer Science and Technology* pp. 634–648 (Elsevier Applied Science, London, 1986).

Williams, J. G., 'Linear fracture mechanics', *Advances in Polymer Science*, **27** (1978) 67–120.

Williams, J. G., *Stress Analysis of Polymers*, 2nd edn (Longman, London, 1981)

Williams, J. G., *Fracture Mechanics of Polymers* (Ellis Horwood, Chichester, 1984).

Thermal Properties

Anderson, D. R. and Acton, R. U., 'Thermal properties', in *Encyclopaedia of Polymer Science and Technology*, vol. 13, pp. 764–788, (Wiley, New York, 1970).

Birley, A. W., and Couzens, D. C. F., 'Thermal properties', in R. M. Ogorkiewicz (Ed.), *Thermoplastics: Properties and Design*, ch. 8 (Wiley, London, 1974).

Brandrup, J. and Immergut, E. H. (Eds), *Polymer Handbook*, 2nd edn (Wiley, New York, 1975).

Hands, D., 'The thermal transport properties of polymers', *Rubber Chem. Technol.*, **50** (1977) 480–522.

Kline, D. E. and Hansen, D., 'Thermal conductivity of polymers', in P. E. Slade, Jr and L. T. Jenkins (Eds), *Thermal Characterization Techniques*, ch. 5 (Dekker, New York, 1970).

4

Electrical and Optical Properties

As polymer materials have developed, their excellent and sometimes outstanding dielectric properties have guaranteed their widespread use as insulants in electrical and electronic engineering. In the nineteenth and early twentieth centuries electrical apparatus relied on wood, cotton sleeving, natural waxes and resins and later ebonite as insulating materials. Today a number of polymers including PTFE, PE, PVC, EP and MF offer an unrivalled combination of cost, ease of processing and electrical performance. These materials have played a most important part in the evolution of electrical components and equipment. Most electrical properties are determined largely by primary chemical structure, and are relatively insensitive to microstructure. In consequence the electrical behaviour of polymers is generally less varied than the mechanical behaviour. The same can be said of the optical properties, which nevertheless govern a variety of engineering end-uses.

4.1 Behaviour in a Steady (d.c.) Electric Field

The electrical properties of a material may be investigated by considering its response to imposed electric fields of various strengths and frequencies, just as the mechanical properties may be defined through the response to static and cyclic stress. We consider first the behaviour of polymers in steady (d.c.) electric fields.

Figure 4.1 shows the range of *volume resistivity* ρ found in electrical engineering materials. Polymers as a class have the very high electrical resistivity characteristic of insulators. To quote resistivities in this way implies that Ohm's law is obeyed and that the conduction current $I = AE/\rho$, where E is the electric field strength and A the cross-sectional area of the material. However, it is

100

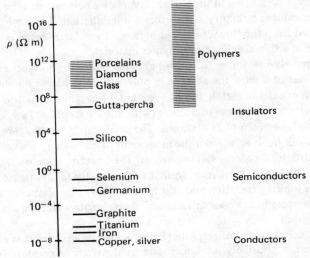

Figure 4.1   Volume resistivity ρ of electrical engineering materials

difficult to measure steady d.c. conduction in high performance insulants such as PTFE, where ρ may exceed $10^{16}\,\Omega$ m. The volume resistivity measured by standard test methods increases steadily with time. One-minute values of ρ are frequently quoted, but curves such as those of figure 4.2 showing how ρ changes with time of electrification present the d.c. conduction behaviour more

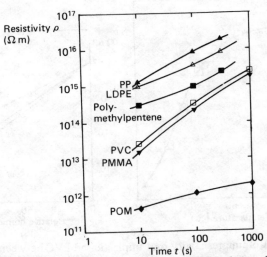

Figure 4.2   Changes in volume resistivity ρ with elapsed time for selected polymers

satisfactorily. Conduction in the surface layers of a polymer material is often sensitive to ambient humidity and surface contamination. The *surface resistivity* is determined from the flow of current between two electrodes in contact with one surface of a thin specimen of polymer material.

The extremely low values of current at typical working voltages implied by the high values of ρ show the absence in polymers of any large number of charge carriers such as exist in metals and to a less extent in semiconductors. The valency electrons in polymer molecules (with a few exceptions) are localised in covalent bonds between pairs of atoms. The small currents which are observed to flow in weak fields arise from the movement of electrically charged species present as structural defects and impurities. The concentration of defects increases as the temperature rises, so that the resistivity falls — figure 4.3(a). Exposure to ionising radiation and absorption of water or plasticiser can also lead to an increase in the concentration of charge carriers with an accompanying increase in conductivity — figure 4.3(b).

The high conductivity of graphite (figure 4.1) demonstrates that polymeric materials are not invariably insulators. The conductivity of graphite (which is a form of pure carbon) arises directly from its chemical and electronic structure. In the solid the carbon atoms lie in parallel stacked layers — *see* figure 1.8(a). Within each layer the C atoms are bonded in hexagonal rings to form a continuous network of primary chemical bonds. However, a quarter of all the valency electrons in such a fused ring structure are not localised between pairs of atoms, but are delocalised, much as they are in metals, and are free to carry a

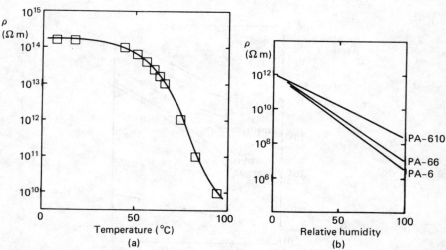

Figure 4.3   (a) Volume resistivity ρ of unplasticised PVC between 0 and 100 °C (all values measured after 60 s electrification). (b) Steady-state volume resistivity ρ of polyamides: dependence on relative humidity at room temperature

conduction current. The conductivity of graphite is much higher parallel to the layers than perpendicular to them because the delocalisation is essentially confined to the individual layers. Carbon fibre and other forms of polymeric carbon with unsaturated structures show similar high conductivity. In sharp contrast diamond, which is another form of pure carbon but which contains only localised electron pair bonds, is a dielectric. The hydrocarbon polymer polyacetylene has some electron delocalisation along the chain and is a semi-conductor with a resistivity of about 10^3 ohm m. It is likely that commercially important engineering polymers with intrinsic conductivity or semiconductor or photoconductor properties will be developed.

In a perfect insulator no steady current flows in a static electric field, but energy is stored in the material as a result of dielectric polarisation. The effect is analogous to the storage of mechanical energy in a perfect elastic material, and arises through the displacement of electric charge. Some polarisation occurs in all materials through small displacements of electrons and nuclei within individual atoms, but larger effects arise if the solid contains permanent dipoles (from polar bonds or asymmetric groups of atoms) which tend to orient themselves in the direction of the externally imposed field. At normal electric field strengths the dielectric polarisation is proportional to the field strength, and we are able to define an important linear property of the material the *relative permittivity* (or dielectric constant), ϵ_r. ϵ_r is the ratio ϵ/ϵ_0 of the permittivity of the material to the permittivity of a vacuum. The permittivity determines the size of the force acting between a pair of electric charges separated by the dielectric material — figure 4.4(a). The capacitance C of a parallel plate capacitor is proportional to the relative permittivity of the medium between the plates. The energy stored by a capacitor charged at voltage V is $\frac{1}{2} CV^2$ and therefore the energy stored by a dielectric in an electric field increases with its permittivity.

The d.c. relative permittivities of selected polymers are listed in table 4.1. The values are determined largely by the nature and arrangement of the bonds in the primary structure. In polymers such as PE, PP and PTFE there is no dipole in the mer because of symmetry. Furthermore both bonding and non-bonding electrons are tightly held and displaced little by external fields. These materials exhibit very little dielectric polarisation, and as a result ϵ_r is low. Polar polymers such as PMMA, PVC and notably PVDF possess higher values of ϵ_r.

The insulating property of any dielectric breaks down in sufficiently strong electric fields. However, in polymers the *dielectric* (or *electric*) *strength* may be as high as 1000 MV/m. An upper limit on dielectric strength is set by the ionisation energies of electrons in covalent bonds within the polymer primary structure. Purely electrical or *intrinsic breakdown* occurs when appreciable numbers of electrons are detached from their parent molecules and accelerate in the electric field to cause secondary ionisation and avalanching. Breakdown of this kind runs its course extremely rapidly and the breakdown voltage does not depend greatly on temperature.

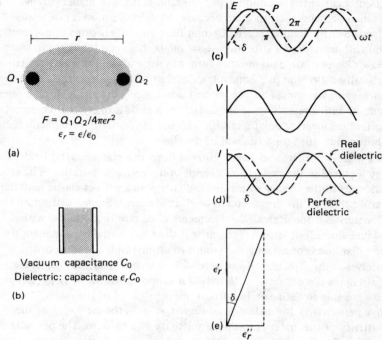

Figure 4.4   Relative permittivity ϵ_r. (a) Screening of electric charges Q_1 and Q_2 by a dielectric of permittivity ϵ. (b) Capacitance of a parallel plate condenser with dielectric. (c)–(e) Response of a dielectric in an alternating electric field. (c) Polarisation response P in a sinusoidal electric field E. (d) Voltage–current relation of a perfect (lossless) and of a real dielectric, showing phase angle δ. (e) Real and imaginary parts of the complex relative permittivity ϵ_r^*

Dielectric breakdown may occur at lower electric field strengths for several reasons. If power dissipated in the dielectric is not lost to the surroundings the rising temperature may bring about *thermal breakdown*. Breakdown voltage in this case depends on heat loss, and hence on the geometry of the specimen and the ambient temperature. In some materials, such as PE above 50 °C, the polymer becomes severely compressed in intense electric fields, and failure may occur through mechanical collapse. Surface contamination (for example by dust and moisture) may lead to local breakdown by *tracking*, a mode of failure in which carbon conducting paths are formed across the surface by localised pyrolysis of polymer. Polymer materials differ greatly in their tendency to fail by tracking, PTFE and PE/PP elastomer (EPM) having particularly good resistance to tracking. Alternatively, discharge through air or vacuum from an electrode to the polymer surface may cause erosion of the material; if the solid polymer contains voids (as in the case of sintered granular PTFE) such discharge erosion may penetrate through the material, and reduce the dielectric strength considerably. Standard

TABLE 4.1
D.C. relative permittivity (dielectric constant)
of selected polymers

	ϵ_r'
PE	2.3
PP	2.3
Polymethylpentene	2.1
POM copolymer	3.8
PMMA	3.8
PVC	3.8
PTFE	2.1
EPDM	3.1
Chlorosulphonated PE (CSM elastomer)	8–10
Urethane elastomer (AU, EU)	9
PVDF	9–13

tests for dielectric strength do exist, but measured breakdown voltages should be regarded as markedly dependent on test and specimen conditions.

4.2 Behaviour in an Alternating (a.c.) Electric Field

In section 3.11 we examined the linear mechanical stress response of a polymer material to a cyclic strain. Stress lags strain by a phase angle δ. We defined a complex compliance $D^* = D' - iD''$, and showed that $D''/D' = \tan \delta$, the loss tangent. The dynamic mechanical behaviour of the material is described by the frequency dependence of D^*.

In a similar way the linear response of a dielectric to an alternating electric field is described by the use of a complex relative permittivity $\epsilon_r^* = \epsilon_r' - i\epsilon_r''$. The ratio ϵ_r''/ϵ_r' is the dielectric loss tangent $\tan \delta$ (known also as the dissipation factor). δ is the phase lag between the electric field and the polarisation of the dielectric – figure 4.4(c). In a perfect dielectric (which behaves as a pure capacitance) the phase angle between current and voltage is $\pi/2$. In real dielectrics, δ is not zero, and current leads voltage by $(\pi/2 - \delta)$. Sin δ is known as the *power factor*. Finally we note that $\epsilon_r'' = \epsilon_r' \tan \delta$ is the *dielectric loss index* of the material.

The a.c. electrical properties of polymers are thus commonly and conveniently expressed in the form of relative permittivity ϵ_r' and loss tangent ϵ_r''/ϵ_r' data as functions of frequency (and temperature). Data may extend from d.c. or low audiofrequency (30 Hz) in half-decade steps to high radiofrequency (1000 MHz) –*see* figure 4.5. In pure homogeneous nonpolar polymers, the polarisation arises

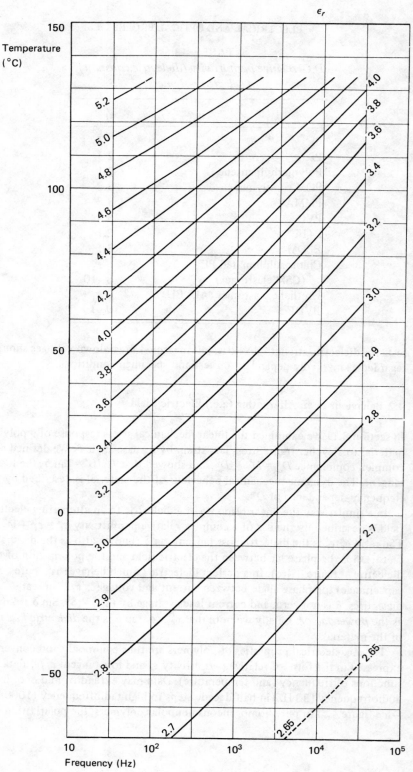

(a)

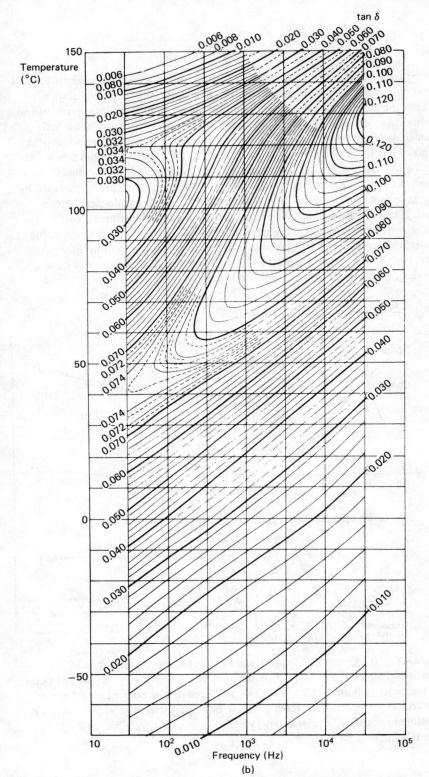

Figure 4.5   Frequency and temperature dependence of (a) permittivity and (b) loss tangent of PMMA (cast sheet, dry) (Data from ICI Ltd)

107

almost entirely from displacements of electrons and nuclei. These redistributions occur extremely rapidly, in picoseconds or less, and the polarisation response follows the alternating electric field without lag up to high radiofrequencies and beyond. As a result polymers of this type, notably PTFE and PE, show little frequency dependence of ϵ_r' and extremely low loss tangents with no loss peaks. In homogeneous PTFE tan δ may be as small as 10^{-5}, and since tan $\delta \approx \delta$ for small δ the loss tangent (or loss angle) is often expressed in microradians. The loss tangents of pure PE and PTFE are so low that they may be sharply increased by very small concentrations of impurities and additives, or by physical heterogeneity. PEs synthesised by different processes may show different loss characteristics depending on the number of CO impurities incorporated in the chain. Sintered granular PTFE shows a loss peak which is absent in material of high crystallinity – figure 4.6(a). PTFE with a high filler content may have a loss tangent as high as 0.1 in damp conditions.

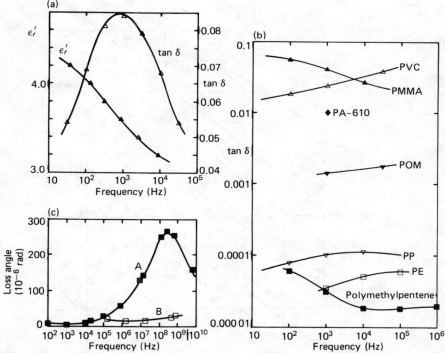

Figure 4.6 (a) Frequency dependence of the relative permittivity and loss tangent of PMMA at 70 °C. (b) Frequency dependence of loss tangents of polar and nonpolar polymers at 20 °C. (c) Dependence of loss tangent on frequency for two PTFE materials: sintered granular PTFE, crystallinity 60 per cent (A); unsintered coagulated dispersion polymer, 93 per cent crystallinity (B) (Data from ICI Ltd)

The loss tangents of polar polymer materials are generally much higher than those of nonpolar polymers — figure 4.6(b). As figure 4.6(c) shows, loss peaks are observed at frequencies corresponding to the relaxation times for dipole reorientation and the positions of these loss peaks can often be correlated with known structural transitions, such as T_g and T_m. The permittivity and loss in the vicinity of a loss peak may be markedly affected by changes in temperature.

The extremely low dielectric loss of PE and PTFE make these outstanding materials for high frequency electronic devices and components. The lossy polymers such as PVC are valuable insulants (for example as cable sheathing) in low frequency applications. At higher frequencies dielectric heating (and the possibility at high voltages of thermal breakdown) may become serious in high loss polymers.

A heated polymer polarised in a very strong electric field and then cooled may retain its polarisation indefinitely. Such a polarised dielectric is the electrostatic counterpart of a magnet, and is known as an *electret*. Polymeric electrets can be prepared with field strengths of 30 kV/cm, showing little loss of strength over periods of several years. Electrets bearing a net electric charge stable for long periods can be made from suitable polymer materials carrying metal foil on one face. Such foil-electrets are used, for example, in microphones.

Finally, we mention the static electrification or triboelectric charging of polymers by contact or friction. The high resistivity of polymers allows large electrostatic charges to accumulate. This has many troublesome consequences (as in textile manufacture and use) and is occasionally hazardous. The study of how triboelectric charging depends on polymer structure is in its infancy, but table 4.2 shows a *triboelectric series* which indicates the sign of the electrostatic

TABLE 4.2
Triboelectric series

Positive end	wool
	PA
	cellulose
	cotton
	silk
	CA
	PMMA
	PVAL
	PETP
	PAN
	PVC
	PVDC
	PE
Negative end	PTFE

charge acquired by contact between pairs of dissimilar polymers. *Antistatic* formulations of polymers have much reduced resistivities so that electrostatic charge may be dissipated by leakage currents.

4.3 Optical Properties

The highest a.c. electrical frequencies we have discussed fall within the short radiowave and microwave regions of the electromagnetic spectrum. The transmission and absorption of radiation by the polymer dielectric are determined by the quantities ϵ_r' and ϵ_r''. Even higher frequencies bring us into the infrared and visible parts of the spectrum, that is, to optical phenomena. In the rest of this chapter we discuss the transmission, scattering and absorption of light by polymer materials.

There is an important underlying continuity in the electrical and optical behaviour of materials; for example, the optical *refractive index n* and the high frequency permittivity are linked in electromagnetic theory by the relation $n = (\epsilon_r')^{1/2}$. At optical frequencies as at lower a.c. frequencies absorption of radiation occurs by irreversible non-radiative loss processes. As we shall see, absorption of optical radiation occurs by excitation of bond vibrations (infrared) and rearrangements of electrons within molecules (visible and ultraviolet). Even so we should not overstress the internal connections between electrical and optical properties, which in practice are usually considered separately and have distinct terminologies. In addition we shall see that *scattering* phenomena play a major part in determining optical properties. Light scattering is a manifestation of optical diffraction processes, which occur prominently in polymer materials having microstructural features of dimensions comparable with optical wavelengths (0.5 μm). Diffraction effects within materials are rarely important at electrical a.c. frequencies because of the much greater associated wavelengths.

4.4 Colour and Infrared Absorption

Very few pure polymers absorb radiation in the visible spectrum, that is, roughly between 380 and 760 nm. Thus most polymers are colourless. The only exceptions are some thermosets and elastomers, including PF, some polyurethanes, epoxies and furan resins, which absorb more or less strongly at the blue end of the spectrum and consequently appear brownish when viewed by transmitted or reflected light. These substances contain alternating double and single covalent bonds or aromatic rings which act as *chromophores*, absorbing light at frequencies corresponding to the excitation energies of bonding electrons. Graphite and the other polymeric carbons which have fused ring structures and which appear intensely black exhibit this absorption throughout the visible spectrum in extreme form. Deliberately coloured polymer materials are

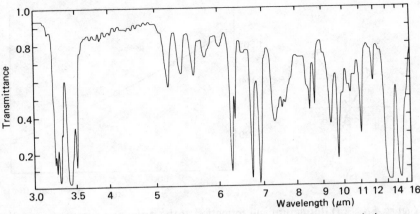

Figure 4.7   Infrared spectrum of polystyrene measured in transmission

invariably produced by incorporating as additives either finely divided coloured solids (pigments) or soluble coloured substances (dyes). The colours of solid polymers are fully defined by their spectra, measured either by transmission or reflection as appropriate. It is often more useful for purposes of specification to refer to colours by using the CIE trichromatic system of colour measurement.

Although many polymers show no absorption of radiation at visible wavelengths, they invariably absorb strongly at certain frequencies in the infrared. The infrared (transmission or reflectance) spectrum of a polymer is determined by its molecular structure. Infrared spectroscopy is one of the most powerful methods of polymer analysis, enabling polymer materials of unknown composition to be identified very rapidly from small samples. Figure 4.7 shows the infrared spectrum of polystyrene PS. The sharply defined absorption peaks correspond to various modes of vibration of individual chemical bonds, and give the analytical chemist detailed information about the primary molecular structure. The infrared spectrum is a fingerprint which allows even closely similar materials such as the various polyamides to be distinguished.

4.5 Refraction

The visual appearance and optical performance of a polymer material depends, apart from colour, on the nature of its surface and its light transmission properties. We follow the recommendations of BS 4618 : Section 5.3 : 1972 and consider in turn *refraction, transparency, gloss* and *light transfer*.

The optical phenomena which occur when a ray of light strikes the plane surface of a colourless transparent material, figure 4.8(a), are governed by the refractive index n. Snell's law relates the angle r to the angle i; Fresnel's equation

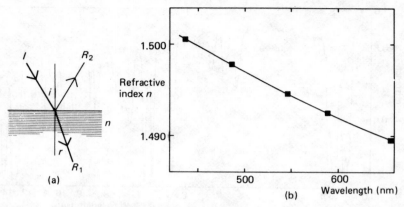

Figure 4.8 (a) Refraction and reflection at the surface of a transparent solid.
(b) Variation with wavelength of the refractive index *n* of PMMA

allows the intensities and polarisations of the refracted and reflected rays R_1 and
R_2 to be calculated from *n* and *i*. *n* varies with wavelength and accurate deter-
minations are made with monochromatic light at specified wavelengths, using
selected lines from the atomic emission spectra of elements contained in
discharge lamps. The wavelength dependence of the refractive index, figure
4.8(b), which leads to refractive dispersion of white light is expressed in terms
of the *reciprocal dispersive power* V_d of the material

$$V_d = \frac{n(d) - 1}{n(F) - n(C)}$$

where d denotes the helium emission line at 587.6 nm; and F and C denote
hydrogen lines at 486.1 and 656.3 nm.

Table 4.3 lists the refractive indices of a number of polymers. More precise
data for polymers which find use as optical materials are given in table 4.4
which includes for comparison two optical glasses.

Refractive index is determined by the extent to which the electronic structure
of the polymer molecules is deformed by the optical frequency electric field of
the radiation. If a material is structurally isotropic, as in the case of unstressed
amorphous polymers, then it is also optically isotropic, and a single refractive
index characterises the refraction behaviour. In crystals and other anisotropic
materials the refractive index takes different values along different principal
axes, and the material is said to be doubly refracting or *birefringent*. Amorphous
materials under deformation develop birefringence as molecules become aligned.
Similar effects arise in flowing polymer melts. The study of birefringence is thus
a very useful method of exploring the microstructural effects of deformation.
One particularly important technical application of stress-induced birefringence in

TABLE 4.3
Refractive index of selected polymers

	Refractive index
PTFE	1.35
PCTFE	1.42–1.43
PVDF	1.42
CA	1.48–1.50
POM	1.48
PMMA	1.49
PP	1.49
PF	1.5–1.7
EP	1.5–1.6
LDPE	1.51
PAN	1.52
NR	1.52
PA	1.53–1.55
HDPE	1.54
PVC	1.54–1.55
UF	1.54–1.56
CR	1.55
PS	1.59
PC	1.59
PVDC	1.60–1.63

TABLE 4.4
*Refractive index $n(d)$ and reciprocal dispersive power V_d
of optical materials*

	$n(d)$	V_d
PS	1.591	31.0
PC	1.586	29.9
PMMA	1.492 6	59.3
Allyl diglycol carbonate	1.498	53.6
Polymethylpentene	1.465 4	44.9
SAN	1.569	35.7
Borosilicate crown glass	1.509 7	64.4
Dense flint glass	1.620 5	36.2

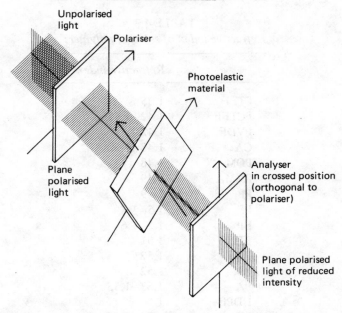

Unpolarised
light

Polariser

Photoelastic
material

Plane
polarised
light

Analyser
in crossed position
(orthogonal to
polariser)

Plane polarised
light of reduced
intensity

Figure 4.9   Principle of photoelastic stress analysis (*see* text)

polymer materials is in *photoelastic stress analysis*, a method of investigating complex stress distributions in engineering components and structures.

Figure 4.9 summarises the principle of a simple photoelastic analysis. The component under examination is modelled in a stress-free polymer material either by moulding or machining. The model component is then loaded and viewed by transmitted light between crossed polaroids. Birefringence induced by the internal stress field generates two components of the refracted ray which travel at different speeds, and may thus produce interference effects. For example, variations in stress within the component give rise to a fringe pattern from which the principal stress lines can be obtained. The value of photoelastic analysis depends almost entirely on the availability of polymeric materials (notably epoxies, polyesters, PC, PS and PMMA) with a suitable combination of mechanical and optical properties. The change Δn in refractive index produced by a stress P defines the *stress-optical coefficient C* of the material by the relation $\Delta n = CP$ (table 4.5).

4.6 Transparency

Colourless polymeric materials range from highly transparent to opaque. Loss of transparency arises from light scattering processes within the material, which distort and attenuate the transmitted image. Figure 4.10(a) shows a light ray I

TABLE 4.5
*Stress-optical coefficient C of commonly used
photoelastic materials*

	$10^{12}\ C(m^2\ N^{-1})$
Selected EP resins	55
PMMA	−4.7
PC	78
Polyurethane elastomer	3500

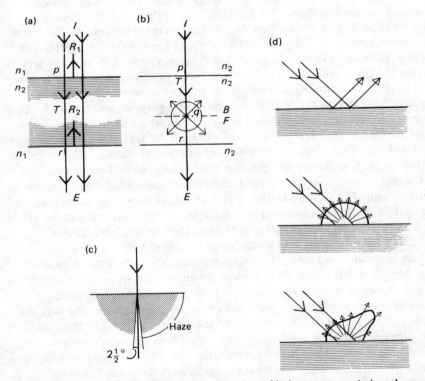

Figure 4.10 (a) Reflection and (b) scattering of light on transmission through a transparent medium. (c) Haze produced by wide-angle forward scattering of light. (d) Gloss: (from the top) specular reflection; perfect diffuse reflection; intermediate behaviour. The polar envelope shows the angular distribution of reflected light intensity

striking a plane surface at right angles. A small fraction of the incident flux is reflected back from the surface at the point of incidence p. The ratio of reflected to transmitted flux is calculated by Fresnel's equation and for normal incidence $\phi(R_1)/\phi(T) = (n_2 - n_1)^2/4n_1n_2$. At an air–polymer interface, with $n_2 = 1.5$, about 4 per cent of the incident flux is reflected. If the air is replaced by a transparent liquid having the same refractive index as the polymer the surface reflection at p (and also at r) is eliminated — figure 4.10(b).

If the material is non-absorbing and optically homogeneous and isotropic the transmitted ray propagates without loss of intensity. In real materials the transmitted ray may lose some intensity as it travels forward as the result of scattering from refractive index inhomogeneities within the material figure 4.10(b). Such scattering is akin to the ray splitting which occurs at the surface at p and which is repeated whenever the ray encounters a change of refractive index δn_2 at points q within the material. The angular distribution of the scattered light is unlikely to be simple, but it is useful to speak of back-scattered and forward-scattered components B and F of the total flux. In figure 4.10(b) we have $\phi(I) = \phi(B) + \phi(F) + \phi(E)$, where E is the undeviated emergent ray. The *direct transmission factor* T of the material is determined as the ratio $\phi(E)/\phi(I)$. For weakly scattering colourless materials T decreases exponentially with the sample thickness l, and therefore $\ln T = -\sigma l$, where σ is the *scattering coefficient*.

The effect of scattering is to reduce the contrast between light and dark parts of the object viewed through the transparent polymer, and thus to produce *haze* or milkiness in the transmitted image. The direct transmission factor, the *total transmitted flux* ϕ_T, i.e. the sum of $\phi(F)$ and $\phi(E)$, and the *forward-scattered fraction* $\phi(F)/\phi_T$ can be determined photometrically.

Besides haze, scattering may produce loss of *clarity* so that distinct features of the object can no longer be distinguished in the transmitted image. Loss of clarity is caused by forward scattering at angles very close to the undeviated beam, since it arises from confusion of light rays diverging from the object at the limits of the angular resolution of the eye. Haze on the other hand is associated with scattering over a wide range of angles. A quantitative measure of haze is obtained in an ASTM standard test from the total forward-scattered flux, *excluding* that within 2½ degrees of the undeviated ray — *see* figure 4.10(c).

In polymer materials heterogeneity of refractive index can arise from differences of density in amorphous and crystalline regions within the polymer itself, or from solid particles incorporated as pigments or fillers, or from voids. The degree of scattering depends strongly on the variation of refractive index ($\pm \delta n_2$) and on the size of the heterogeneities. The most efficient scattering occurs when the scattering centres are comparable in dimensions with the wavelength of light. The scattering of light in pigmented polymer films has been extensively studied in connection with the optical properties of paints (*see* section 6.8). Crystalline polymers are normally translucent or opaque unless it happens (as in polymethylpentene) that there is little difference in the refractive indices of crystalline and amorphous regions or (as in cellulose triacetate) that

TABLE 4.6
Selected optical properties of some engineering polymers

	PP	LDPE (rapidly cooled film)	Polymethyl-pentene (modified)	PA (nylon 66)	PMMA	PVC*	PTFE
Refractive index n (546.1 nm)	1.505	1.516	1.472	–	1.495	–	–
Refractive index n (589.3 nm)	–	–	1.465	1.448	–	1.542	1.325
Temperature coefficient of refractive index ($10^4 \, dn/dT$)(K^{-1})	–	–2.4	–2.4	–	–1.0	–1.1	–
Direct transmission factor T(%) (normalised to 1 mm sample thickness)	11	45	99	0	99	94	0
Scattering coefficient $10^{-2} \, \sigma$ (m^{-1})	21.5	8.3	0.11	100	0.01	0.6	80

* Optical properties vary widely between grades.

the spherulites are exceptionally small. It is frequently possible to enhance the transparency of polymer materials by assisted nucleation or by rapid cooling from the melt, both means of decreasing the spherulite size. Stretching is also an effective way of increasing transparency, since spherulites are transformed into oriented fibrils which scatter light less effectively.

4.7 Gloss

'Gloss' is a term applied to the surface optical character of a material, whether transparent or opaque. A perfect mirror-like surface, a specular reflector, shows one extreme of behaviour — figure 4.10(d). At the other extreme, a highly scattering surface (a perfect diffuse reflector) reflects light equally in all directions at all angles of incidence.

The gloss of real surfaces is most completely described by the angular intensity distribution of reflected light for one or more angles of incidence. A somewhat simpler index of surface appearance is the *direct reflection factor*, the ratio of the flux reflected at the specular angle to the incident flux, for angles of incidence from 0 to 90 degrees.

Table 4.6 collates optical data on a number of engineering polymers.

4.8 Light Transfer

Strongly scattering translucent materials (optical 'diffusers') are unable to transmit useful images but nevertheless may transmit appreciable amounts of diffuse light. Their optical characteristics are most simply expressed through the *total transmission factor* T (or *luminous transmittance*) for normal incidence, which is defined as $[\phi(F) + \phi(E)]/\phi(I)$.

Suggestions for Reading

Electrical Properties

ASTM D149–81, *Dielectric breakdown voltage and dielectric strength of solid electrical insulating materials at commercial power frequencies* (American Society for Testing and Materials, Philadelphia, Pa.).

ASTM D150–81, *A-C loss characteristics and permittivity (dielectric constant) of solid electrical insulating materials* (American Society for Testing and Materials, Philadelphia, Pa.).

ASTM D257–78, *D-C resistance or conductance of insulating materials* (American Society for Testing and Materials, Philadelphia, Pa.).

Baird, M. E., *Electrical Properties of Polymeric Materials* (The Plastics Institute, London, 1973).

Block, H., 'The nature and application of electrical phenomena in polymers', *Appl. Polymer Sci.*, **33** (1979) 93–167.

Blythe, A. R., *Electrical Properties of Polymers* (Cambridge University Press, 1979).

Brown, R. P. (Ed.), 'Electrical properties', in *Handbook of Plastics Test Methods*, 2nd edn (Godwin, London, 1981).

BS 4618: Part 2: 1970, *Electrical properties.*

Buckingham, K. A., 'Electrical properties', in R. M. Ogorkiewicz (Ed.), *Thermoplastics: Properties and Design*, ch 7 (Wiley, London, 1974).

Goosey, M. T. (Ed.), *Plastics for Electronics* (Applied Science, London, 1985).

Harper, C. A., 'Electrical applications', in *Encyclopaedia of Polymer Science and Technology*, vol. 5, pp. 482–528 (Wiley, New York, 1966).

Harrop, P., *Dielectrics* (Butterworths, London, 1972).

Ku, C. C. and Liepins, R., *Electrical Properties of Polymers* (Hanser, Munich, 1987).

Mathes, K. N., 'Electrical properties', in *Encyclopaedia of Polymer Science and Engineering*, 2nd edn, vol. 5, pp. 507–587 (Wiley, New York, 1986).

Norman, R. H., *Conductive Rubbers and Plastics* (Elsevier, Amsterdam, 1970).

O'Dwyer, J. J., *The Theory of Electrical Conduction and Breakdown in Solid Dielectrics* (Oxford University Press, 1973).

Seanor, Donald A., 'Electrical properties of polymers', in A. D. Jenkins (Ed.), *Polymer Science*, vol. 2, ch. 17 (North-Holland, Amsterdam, 1972).

Seanor, D. A. (Ed.), *Electrical Properties of Polymers* (Academic Press, New York, 1982).

Sillars, R. W., *Electrical Insulating Materials and their Application*, IEE Monograph Series 14 (Peter Peregrinus, Stevenage, for The Institution of Electrical Engineers, 1973).

Solymar, L. and Walsh, D., *Lectures on the Electrical Properties of Materials*, 4th edn (Oxford University Press, 1988).

Williams, M. W., 'Dependence of triboelectric charging of polymers on their chemical compositions', *Rev. Macromol. Chem.*, **14B** (1976) 251–265.

Wintle, H. J., 'Theory of electrical conductivity of polymers', in M. Dole (Ed.), *The Radiation Chemistry of Macromolecules*, vol. 1, pp. 109–126 (Academic Press, New York, 1972).

Optical Properties

ASTM D523-85, *Specular gloss* (American Society for Testing and Materials, Philadelphia, Pa.).

ASTM D1003–61, *Haze and luminous transmittance of transparent plastics* (American Society for Testing and Materials, Philadelphia, Pa.).

Brown, R. P. (Ed.), 'Optical properties', in *Handbook of Plastics Test Methods*, 2nd edn (Godwin, London, for the Plastics Institute, 1981).

BS 4618: Section 5.3 : 1972, *Recommendations for the presentation of plastics design data: optical properties.*

Hunter, R. S. and Boor, L., 'Tests for surface appearance of plastics', in John V. Schmitz (Ed.), *Testing of Polymers*, vol. 2 (Interscience, New York, 1966).

Kuske, A. and Robertson, G., *Photoelastic Stress Analysis* (Wiley–Interscience, New York, 1974).

Meeten, G. H. (Ed.), *Optical Properties of Polymers* (Elsevier Applied Science, London, 1986).

Meinecke, E., 'Optical properties: survey' in *Encyclopaedia of Polymer Science and Technology*, vol. 9, pp. 525–551 (Wiley, New York, 1968).

Redner, S., 'Optical properties: photoelasticity', in *Encyclopaedia of Polymer Science and Technology*, vol. 9, pp. 590–610 (Wiley, New York, 1968).

Ross, G. and Birley, A. W., 'Optical properties of polymeric materials and their measurement', *J. Phys. D.: Appl. Phys.*, 6 (1973) 795–808.

Ross, G. and Birley, A. W., 'Optical properties', in R. M. Ogorkiewicz (Ed.), *Thermoplastics: Properties and Design* (Wiley, London, 1974).

Wright, W. D., *Measurement of Colour*, 3rd edn (Hilger & Watts, London, 1964).

5

Chemical Properties

Not the least remarkable and useful of the properties of polyethylene is its chemical inertness. PE is unaffected by prolonged contact with those strong acids (including hydrogen fluoride) which attack many metals, highly caustic alkalis which rapidly damage glass, as well as most organic substances. Yet the pure polymer weathers rapidly when exposed outdoors to the atmosphere, burns easily and may be affected by common detergents. These properties are all aspects of the *chemical* behaviour of the material, the subject of this chapter. The contrast between the inertness of PE and some other polymers towards many aggressive reagents, including solvents, and its sensitivity to atmospheric degradation warn us to expect some complexity in the chemical properties of polymers. We shall discuss a number of agents of chemical change which are of engineering importance for polymer materials. These are solvents, environmental agents (producing weathering and ageing), fire, radiation, and biological organisms.

5.1 Solubility and Solution Properties

The action of solvents stands apart somewhat in this list of themes because it does not produce permanent chemical change in the polymer material. However, solvents produce profound physical changes in the polymer and these changes occur as a result of the chemical nature of both the solvent and the polymer. The technical importance of the solubility properties of polymers is enormous, and has no real counterpart in the technology of other major classes of engineering materials. Four examples of technical processes depending on solvent action illustrate this.

(1) Many polymers can be joined by solvent welding, or by the use of adhesives consisting of dissolved polymer.

(2) Synthetic fibres can be produced by continuous evaporation of solvent from jets of polymer solution.

(3) Polymer films and sheet are produced by solvent casting. Similarly film formation in paint-drying frequently occurs by solvent evaporation.

(4) In two-phase polymer blends, differences in solubility between the two phases are used to produce filtration membranes and to modify surface properties.

Modern approaches to understanding solubility require us to consider the free energy differences between the two states A and B of figure 5.1. If A has a lower free energy than B, the solid substance does not pass spontaneously into solution – it is insoluble. We recall that the free energy is a composite quantity comprising an enthalpy term (which depends on the strength of the van der Waals' forces acting between molecules) and an entropy term (depending on the degree of disorder of the molecules and their constituent parts). If the forces between the solvent molecules and the polymer molecules are stronger than those between adjacent polymer molecules in the solid, then solubility is favoured. The inner cohesion of the polymer solid increases with the degree of crystallinity as the efficiency of molecular packing assists intermolecular interactions. All other things being equal, we may expect crystalline polymers to be less easily dissolved than amorphous. Furthermore the old practical rule *like dissolves like* carries a seed of truth, for at least it is the case that polar solvents (water, alcohols, ketones) are likely to be effective in penetrating and perhaps ultimately dissolving polar polymers.

Flexible polymer chains are more or less firmly held in the solid state but acquire considerable configurational freedom on passing into solution. This is reflected in an increase in entropy on dissolving, a factor which promotes polymer solubility. However, the gain in entropy is not so great for polymers as for substances composed of small independent molecules. As a result many polymers will dissolve in only a few well-matched solvents, and polymers solutions exhibit an unusual variety of phase separation (demixing), partial miscibility and fractionation effects.

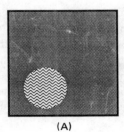 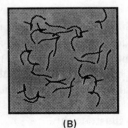

(A)   (B)

Figure 5.1   Polymer solubility. The solid polymer dissolves in the solvent only if state B has a lower free energy than state A

One approach to predicting polymer solubility has been widely applied for technological ends. This in essence assumes that the enthalpic term is dominant. A *solubility parameter* δ is defined for both solvent and polymer (table 5.1). δ is a measure of the strength of intermolecular cohesion in the pure solvent or in the pure polymer. For the solvent δ is calculated from the energy of vaporisation. For the polymer it is obtained indirectly, and may be estimated from the primary structure of the chain. The significance of the solubility parameter is that *polymers are soluble only in solvents of similar* δ. Thus, for example, natural rubber (δ = 8.3) is dissolved by toluene (δ = 8.9) and carbon tetrachloride (δ = 8.6) but not by ethanol (δ = 12.7). Unfortunately, this general rule only holds if strong polymer–solvent interactions are absent. It is unlikely that any simple predictive method can provide for the whole range of solubility behaviour.

If a solvent S dissolves a polymer P it is invariably found that the polymer dissolves the solvent also. Frequently there are limits to the amount of P which S will dissolve and the amount of S which can be absorbed by P. These limits define the miscibility range of the two substances. Sometimes the polymer and solvent are sufficiently well matched that they are miscible in all proportions. The absorption of small amounts of S by P produces changes in physical, especially mechanical, properties. Controlled use of such *plasticisation* is an important aspect of polymer technology and the matter is taken up in chapter 6. We note in passing, however, that such external plasticisers are free to migrate within and permeate through the polymer material. In the following section we discuss the more important topic of the permeability of polymer materials to gases and vapours.

Crosslinked polymers cannot dissolve in the sense we have described, for individual polymer chains cannot be detached from each other. Such materials show extensive swelling when they come into contact with *compatible* solvents. Such effects are often important in the performance of elastomer components such as seals and hoses.

An especially interesting property of the polymer solution is its viscosity. In dilute solution, the individual polymer chains are well separated; the increase in viscosity (relative to the viscosity of the pure solvent) therefore reflects the hydrodynamic friction caused by the presence of isolated polymer molecules in the flowing solution. As might be expected, this is directly related to the size of the polymer molecule and thus to the molar mass or chain length. The viscosity of dilute solutions of many linear polymers obeys the Mark–Houwink law

$$\left[\left(\frac{\eta}{\eta_0} - 1\right)\bigg/c\right]_{c=0} = [\eta] = KM^a \tag{5.1}$$

where η is the solution viscosity, η_0 the solvent viscosity, and M the molar mass. The quantity $[\eta]$ is the *limiting viscosity number* (or *intrinsic viscosity*), obtained as the limiting slope of $(\eta/\eta_0 - 1)/c$ as the polymer concentration c approaches zero. The Mark–Houwink parameters K and a depend on the polymer and the solvent. This relation is widely used to determine the molar mass of polymer samples since the viscosity is easily measured, usually in a simple U-tube capillary viscometer (*see*, for example, ASTM method D2857). The parameters K and a provide information on the polymer–solvent interaction and on the mean end-to-end separation of the polymer chain in solution (represented in figure 1.6 by the quantity r_p). In a 'good' solvent, the polymer molecule adopts a relatively open random chain conformation, the distance r_p being somewhat greater than the value r_{p0} calculated for the freely jointed chain; when the polymer–solvent interaction is less favourable, the polymer adopts a tighter conformation, reducing its mean radius. The tightest coiling occurs in a solvent in which the polymer will only just dissolve. In this case, the polymer chain tends to coil back upon itself to maximise the number of contacts between different parts of the chain and to minimise molecular contact with the solvent. Further reduction in the quality of the solvent produces precipitation of the polymer as separate chain molecules aggregate together. Solvents in which r_p has the unperturbed value r_{p0} are called Flory or theta solvents. In practice, these are poor solvents and the polymer chain cannot shrink its end-to-end separation much below r_{p0} without precipitating from solution. (Since polymers dissolve in most solvents over a very limited range of temperatures, there is also a theta temperature in any solvent.) In a theta solvent, the Mark–Houwink parameter $a = 0.5$; in better solvents, a is larger, rising to a maximum of about 1.0.

The separation of individual polymer molecules which occurs in dilute solution in good solvents can be exploited in other ways to determine molar mass (or chain length) distributions. The angular intensity variation of light scattered from a dilute polymer solution provides a direct measure of the molar mass distribution of the polymer. In *size exclusion chromatography*, the polymer molecules flowing through a chromatograph column packed with porous solid are retained to an extent which depends on their ability to enter the pores of the packing. This subtle effect causes a separation of molecules according to molecular size (and hence molar mass). An example of a molar mass distribution obtained by this technique (also known as gel permeation chromatography) was shown in figure 1.5.

We have emphasised the ability of solvents to separate polymer molecules and to transform solid materials into a fluid state for various practical purposes. We end this section with a reference to the ability of some polymers in solution to achieve precisely the opposite: namely, to convert the solvent into a solid or semi-solid material. The process of *gelation*, in which small quantities of dissolved polymers link together to form a continuous network throughout the solution, is both useful and scientifically interesting.

An aqueous gel is a random polymer network structure within which substantial amounts of water are retained. Gels are usually formed from aqueous solutions of linear polymers which crosslink by chemical reaction or physical entanglement. As crosslinking develops, the solution viscosity rises rapidly, becoming infinite at the *gel-point* (or the *sol–gel transition*). Several water-soluble polymers can form hydrogels at concentrations as low as 1 per cent by weight of polymer. The properties of chemically crosslinked aqueous gels (particularly of polyacrylamide and acrylamide–acrylic acid copolymers) have been investigated by Tanaka whose work has revealed a variety of striking physicochemical effects. For example, aqueous gel samples exposed to mixed acetone/water solvents or to salt solutions may collapse dramatically at critical acetone or salt concentrations, shrinking in volume by a factor of as much as 100 (figure 5.2). The theory developed to explain these transitions is in essence an extension of the theory of rubber elasticity to take account of the interionic forces between the aqueous solution and the ionic groups attached to the gel network. The phase transitions which are observed arise when the gel is unable to find a stable balance between the osmotic, electrostatic and rubber–elastic forces acting within it.

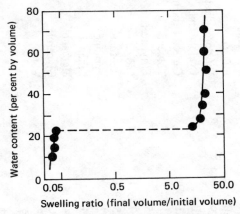

Figure 5.2 Volume of a crosslinked acrylamide/acrylic acid polymer gel in acetone/water mixtures, showing the sudden and very large shrinkage transition at a well-defined acetone concentration (after Tanaka)

5.2 Permeability

Many engineering uses of plastics and rubbers depend on their often excellent barrier properties. The rate at which gases and vapours permeate polymer films depends on the partial pressure gradient across the film (figure 5.3). The following relation generally holds at least approximately (at a given temperature)

$$\text{Flux/Area} = P(p_1 - p_2)/l \qquad (5.2)$$

where the *permeability coefficient* P depends on the polymer material and the permeating substance, partial pressure p. If the steady-state diffusion of permeant within the film obeys Fick's law then

$$\text{Flux/Area} = -D(\partial c/\partial x) \qquad (5.3)$$

TABLE 5.1

Solubility parameter δ of polymers and solvents

Polymers	δ (cal$^{1/2}$ cm$^{-3/2}$)	Solvents	δ (cal$^{1/2}$ cm$^{-3/2}$)
PTFE	6.2	n-Pentane	6.3
Poly(dimethyl siloxane)	7.3	n-Hexane	7.3
PE	8.0	n-Octane	7.6
PP	7.9	Diisopropylketone	8.0
PIB	8.0	Cyclohexane	8.2
SBR	8.1–8.6	Carbon tetrachloride	8.6
Natural rubber	8.1	Toluene	8.9
BR	8.5	Ethyl acetate	9.1
Polysulphide rubber	9.0–9.4	Dioxane	9.9
PS	8.5–9.7	Acetone	10.0
CR	9.2	Pyridine	10.9
PVAC	9.4	Ethanol	12.7
PMMA	9.2	Methanol	14.5
PVC	9.6	Glycerol	16.5
PVDC	9.8	Water	23.4
PETP	10.7		
Cellulose acetate	11.4		
EP	11.0		
POM	11.1		
Nylon 66	13.6		
PAN	15.4		

Note: 1 cal = 4.16 J.

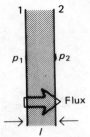

Figure 5.3 Permeation of gases and vapour through a polymer under the action of a partial pressure gradient $(p_1 - p_2)/l$

where D is the *diffusivity*. If D is independent of c then

$$\text{Flux/Area} = -D(c_1 - c_2)/l \tag{5.4}$$

where c_1 and c_2 are the concentrations of the diffusing substance at the surfaces 1 and 2.

If in addition the gas or vapour dissolves in the film according to Henry's law

$$c = Sp \tag{5.5}$$

(S is the solubility). It follows that

$$P = DS \tag{5.6}$$

Thus the observed permeabilities depend on the interplay of diffusion and solubility parameters. Table 5.2 lists permeability coefficients, diffusivities and solubilities of a number of gases and vapours in common polymers. The atmospheric gases generally have relatively low solubilities and obey Henry's law, but diffuse freely through amorphous regions of solid polymer. Vapours of organic substances with δ values similar to the polymer have relatively high solubilities with deviations from Henry's law, but diffuse more slowly. The slower diffusion is a result of larger molecular sizes and often strong interactions with polymer chains. Permeability generally increases with increasing temperature; and for a particular polymer decreases as crystallinity increases (figure 5.4).

Important *high barrier* polymers with low permeabilities to both gases and water include PVDC and copolymers, acrylonitrile–styrene copolymers, EP resins, PVDF, PETP and PVC. PUR, PS and common elastomers have poor barrier properties. PVAL and regenerated cellulose (cellophane) have very low resistance to the passage of water but are good gas barriers when dry. Conversely, PE, PP and PTFE are good water barriers, but have quite high permeabilities to permanent gases. Table 5.3 shows the importance of primary structure in controlling permeability.

TABLE 5.2

Permeability of common polymers to gases and vapours, together with corresponding diffusivity and solubility data at 25 °C

Polymer	Permeant	$10^{15} P$ (kg m^{-1} kPa^{-1} s^{-1})	$10^{12} D$ (m^2 s^{-1})	$10^3 S$ (kg m^{-3} kPa^{-1})
PA 6 (Nylon 6)	Nitrogen	0.023	0.025	0.94
PETP	Nitrogen	0.063	0.13	0.48
PVC	Carbon dioxide	0.52	0.21	2.5
PCTFE–PVDF copolymer	Nitrogen	0.85	0.36	2.4
PIB	Nitrogen	3.1	4.5	0.69
	Carbon dioxide	77	5.8	13
CR	Nitrogen	11	25	0.44
	Carbon dioxide	300	24	16
NR	Nitrogen	76	110	0.69
	Carbon dioxide	1900	110	18
	n-Propane	2500	21	120
HDPE	Helium (30 °C)	1.9	360	0.005 4
	Oxygen (30 °C)	5.4	22	0.25
	Nitrogen (30 °C)	1.7	12	0.14
	Carbon dioxide (30 °C)	31	16	2.0
LDPE	Isobutene (30 °C)	680	4.7	140
	n-Hexane (30 °C)	6200	2.5	2500
	Water	540	23	24

Note: 1 kPa = (1/101.3) atm ≈ 0.01 atm.

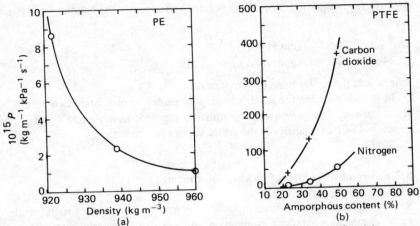

Figure 5.4 (a) Permeability P to nitrogen of PE films of different densities (and degrees of crystallinity), measured at 25 °C. (b) Permeability P to carbon dioxide and nitrogen of PTFE of different crystallinities, measured at 30 °C

A variety of units has been used in reporting permeabilities; among the commonest are

cm^3 (at STP) m^{-2} day^{-1} atm^{-1} (film or sample thickness l stated)
cm^3 (at STP) cm^{-1} (cm Hg)$^{-1}$ s^{-1}
g cm^{-2} s^{-1} (cm Hg)$^{-1}$ (l stated)

In the SI system the quantity D has the units m^2 s^{-1} and the Henry's law solubility S usually has the units kg m^{-3} kPa^{-1}. The units of P then follow

TABLE 5.3
Effect of primary structure on oxygen permeability
P of carbon chain polymers $\{CH_2 - CHX\}$
(adapted from Nemphos *et al.*)

$- X$	$10^{15} P$ (kg m^{-1} kPa^{-1} s^{-1})
$-$ OH	0.000 6
$-$ CN	0.002
$-$ Cl	0.51
$-$ F	0.96
$-$ CH_3	9.6
$-$ C_6H_5	27
$-$ H	31

directly from equation 5.5, that is, kg m^{-1} kPa^{-1} s^{-1}. It follows from the gas laws that

$$cm^3 \text{ (at STP) } cm^{-1} \text{ (cm Hg)}^{-1} s^{-1} \equiv$$
$$(3.346\ 3 \times 10^{-3}\ \text{M/kg mol}^{-1})\ kg\ m^{-1}\ kPa^{-1}\ s^{-1}$$

where M is the molar mass of the permeant.

In British practice the water vapour *permeance* of membranes (whose thickness may not be known) is commonly reported in units g MN^{-1} s^{-1}. The reciprocal of this quantity is the water vapour *resistance*.

5.3 Environmental Stress Cracking and Crazing

Some polymers *when stressed* are adversely affected by contact with certain active chemical substances. Aqueous solutions of surface-active compounds such as detergents can produce slow, largely brittle cracking in stressed PEs. The term *environmental stress cracking* (ESC) was introduced to denote this particular type of failure. ESC is now used more generally to describe the promotion of slow brittle failure in stressed polymers by organic substances, as for example the cracking of PVC gas pipes by certain hydrocarbon impurities. Organic liquids and gases can also promote the formation of fine networks of voids (*crazes*) in amorphous polymers, such as PS. Crazed material retains considerable strength, but it is usually unsightly, and its appearance may precede crack formation. In both ESC and crazing, damage arises from the conjoint action of a substance in the environment and stress (either externally imposed or arising internally from processing). Direct chemical attack on the primary bonds of the polymer chain is not involved. In both cases it appears that the promoting substance is adsorbed or locally dissolved at defects so as to assist failure, perhaps by modification of the surface energy or by plasticisation of highly stressed material at the crack tip. The susceptibility of a polymer to ESC failure depends on structural factors. For example, the resistance of PE to stress cracking varies considerably with molar mass and melt flow index, crystallinity, and density.

5.4 Chemical Attack

The effects of solvents on polymer molecules are usually physical rather than chemical. The primary polymer chain remains intact, and the polymer can be recovered by evaporation of the solvent. Of course the microstructure (morphology) of the recovered polymer solid may be quite different from that of the original solid.

In addition, as organic (carbon-based) compounds polymers are attacked by a number of chemical substances which produce irreversible chemical changes in the chain. These chemical reactions are determined principally by the chemical structure of the polymer. For example, sulphuric acid H_2SO_4 reacts with phenyl groups. Thus PS may be sulphonated

Similarly PE may be (partially) chlorinated in the same manner as short-chain paraffins

$$+CH_2+ \xrightarrow{Cl_2} +CHCl+$$

Those common polymers which are paraffinic in structure exhibit the relative chemical inertness of their class. They are attacked only by strong chemical reagents. However, other polymers (condensation polymers particularly) are susceptible to chemical attack under milder conditions. Manufacturers of commercial polymers provide information on the resistance of their products to a wide range of chemical substances likely (and often unlikely!) to be encountered in engineering applications. Chemical reactions between polymers and other substances may sometimes produce useful modifications of properties, and in fact both the chlorination of PE (and also of PVC and of natural rubber) and the sulphonation of PS yield commercial polymer materials.

Polymers (in common with all other organic compounds) are vulnerable to *oxidation*. Unstabilised polymers undergo slow ageing in contact with air. Outdoor conditions in particular are very aggressive, for oxidation reactions are frequently strongly promoted by the ultraviolet radiation present in sunlight. Degradation of the primary chain occurs more rapidly the higher the temperature, and heat is an agent of polymer deterioration both in the presence and in the absence of oxygen. These various processes are classified in table 5.4.

TABLE 5.4
Major agents and modes of polymer degradation

Oxygen	Moderate temperature	**(Thermal) Oxidation**
	High temperature	**Combustion**
Oxygen	Ultraviolet radiation	**Photo-oxidation**
Water		**Hydrolysis**
Atmospheric oxygen, water, natural light	Moderate temperature	**Weathering, atmospheric degradation**
Heat alone		**Pyrolysis**
Ionising radiation		**Radiolysis**

The oxidation of polymer materials involves free radical chain reactions (*see* section 1.12). Even substances such as PE which are expected to be inert at moderate temperatures in the absence of light react quite rapidly with oxygen unless stabilised. It is not likely that direct attack of oxygen on the strong C—H or C—C bonds occurs under such mild conditions, and it appears that the reaction is initiated by hydroperoxide impurities incorporated during synthesis and processing

$$R\text{—}O\text{—}O\text{—}H \longrightarrow RO\cdot + \cdot OH \qquad (i)$$

and
$$R\text{—}O\text{—}O\text{—}H \longrightarrow RO_2\cdot + \cdot H \qquad (ii)$$

The $RO_2\cdot$ peroxy radicals formed are sufficiently reactive to attack some primary CH bonds of the chain

$$\text{+CH}_2\text{+} + RO_2\cdot \longrightarrow \text{+}\overset{\cdot}{C}H\text{+} + ROOH \qquad (iii)$$

and then very rapidly

$$\text{+}\overset{\cdot}{C}H\text{+} + O_2 \longrightarrow \overset{\overset{\displaystyle O\text{—}\overset{\cdot}{O}}{|}}{\text{+CH+}} \qquad (iv)$$

The peroxy radical is thus reformed and can attack another CH bond. The action of this radical continues (the chain reaction *propagates*) until a termination reaction intervenes. Then

$$\left.\begin{array}{l} 2RO_2\cdot \\ 2R\cdot \\ RO_2\cdot + R\cdot \end{array}\right\} \longrightarrow \text{non radical products} \qquad (v)$$

or

or

Generally the rate of oxidation, after an initial period during which hydroperoxides accumulate, is determined by process (iii) above. Consequently the observed rate of deterioration may be considerably increased by the presence of a small number of easily removed H atoms at irregularities in the chain. For example it is known that a hydrogen atom is less easily removed from a location such as I (secondary H atom) than from II (tertiary H atom).

(I)   (II)

Thus in PE the solitary hydrogen atom directly bonded to the chain at branch points is especially vulnerable. It is found that the rate of oxidation is closely related to the degree of chain branching. The presence of a double bond activates the adjacent H atom; thus saturation (the absence of double bonds) improves oxidation resistance. Because of oxygen permeability differences (*see* section 5.2) crystalline forms of polymers are much more resistant to oxidation than amorphous forms. The C—F bond is much more resistant to free radical cleavage than the methylene CH, and PTFE shows exceptional resistance to thermal oxidation.

It is possible to increase greatly the oxidation resistance of polymers by incorporating small amounts of additives which interfere with the chain mechanism outlined above. Such stabilisers are discussed at greater length in chapter 6.

Both natural rubber and synthetic elastomers such as SBR are attacked by atmospheric ozone. If the materials are under stress, the degradation leads to a characteristic form of deterioration known as *ozone cracking*, in which cracks appear perpendicular to the stress. The ozone content of the atmosphere is usually about 0.01 mg kg^{-1} (one part in 10^8), although in polluted air the concentration may be as much as 100 times greater. Even at such low ozone levels as these, rubbers require stabilisation against ozone attack if acceptable service lifetimes are to be achieved. The O_3 molecule attacks C=C bonds (not a free radical reaction) and can cause scission of the main chain

Natural rubber NR

5.5 Thermal Stability

The oxidation reactions just described proceed more rapidly at higher temperatures. However, even if we can stabilise a polymer against thermal oxidation, the *thermal stability* of most polymers is limited. By comparison with ceramics and metals the useful temperature span of polymer materials is restricted. Table 5.5 indicates the practical maximum service temperature of a number of engineering materials. Of course the limit is often set by deteriorating physical performance rather than by thermal instability of the chemical structure. Nevertheless, heat alone brings about degradation of polymer chains, frequently by main chain scission with the formation of radicals

In some cases, such as PMMA and PTFE, it is the bond between the terminal mer and the rest of the chain which breaks. This unzipping reaction releases

TABLE 5.5
Upper service temperatures of selected materials

Metals and ceramics	(°C)
Plain carbon steels	550
Austenitic stainless steels	800
Nickel chromium alloys	900
Cobalt alloys	950
Molybdenum	1000
Niobium	1150
Silica	1000
Borosilicate glass	450
Asbestos	600
Portland cement concrete	300
Aluminosilicate firebrick	1500
Boron nitride	2000
Graphite	3000

Polymers	Intermittent use	Continuous use
Polyethersulphone	250	190
Polyurethane elastomer	120	80
PA 66	180	110
PMMA	–	80
Acetal copolymer	105	80
PTFE	260	–
EPDM elastomer	180	150
NR	100	80
Fluorocarbon elastomer	260	200
Silicone elastomer	300	–

Note: Data for polymers are based on manufacturers' recommendations for particular materials; precise limits depend on service conditions and performance requirements.

monomer in high yield. (In the hot conditions prevailing in pyrolysis the monomer is volatile and the material loses weight: *see also* section 5.9.)

In the case of other polymers, such as the polyolefins, scission occurs at random locations on the chain and the monomer yield is extremely small. The degradation reduces the chain length (molar mass) and secondary reactions may produce more or less complex mixtures of volatile degradation products.

The thermal degradation of PVC occurs by elimination of HCl gas without scission of the main chain; at 250 °C and above the reaction occurs rapidly and almost completely

$$\text{(CH}_2\text{-CHCl)} \longrightarrow \text{(CH=CH)} + \text{HCl}$$

The reaction is accompanied by embrittlement and by severe discoloration of the material, arising from light absorption by the alternating C=C and C—C bonds of the main chain.

At temperatures above about 400 °C, the rate of degradation of the common polymers is rapid, with pyrolysis largely complete in a few minutes. At such temperatures, the pyrolysis products may be entirely volatile (as when monomer yield is high). In other cases (notably crosslinked polymers and some thermoplastics such as PVC and PAN) much of the main chain carbon is incorporated in a non-volatile char. Figure 5.5 shows the weight loss curve of a laboratory sample of PAN heated at a rate of 5 °C/min. The polymer is stable up to about 250 °C, at which temperature 20 per cent of the material volatilises rapidly with loss of an organic nitrogen compound. Further degradation occurs steadily up to about 480 °C. At this temperature, the sample is a black powder in which nitrogen and carbon atoms have become incorporated in a thermally stable ring structure (*see also* section 6.9).

Such carbonaceous residues, which sometimes contain mineral fillers from the original polymer material, resist much higher temperatures than the parent substance, especially in a non-oxidising atmosphere. Surface chars have low thermal conductivities and may protect underlying material from the effects of heat for a long period. The good performance of timber structures in building fires is explained in this way. Nylon—phenolic (PA—PF) blends are used as ablative heat shields on space vehicles during re-entry, when surface temperatures of several thousand degrees are attained.

Some degree of stabilisation against thermal degradation at modest temperatures (say below 200 °C) is possible through the use of additives. These are important in preventing degradation during polymer processing (for example

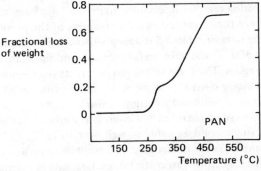

Figure 5.5 Pyrolysis of PAN: loss of weight at a heating rate of 5°C/min

of PVC). The melt temperatures may be sufficiently high for some degradation to occur in unstabilised materials.

A method devised by Underwriters Laboratories is widely used to assess the *thermal endurance* of polymer materials. Put simply, the method determines the maximum temperature at which a physical property retains 50 per cent of its initial value for a period of 60 000 hours. This temperature is called — somewhat curiously — the *relative thermal index* of the material. The test method is based on a statistical analysis of how one or more physical properties (such as impact strength or electrical resistivity) vary with time at different temperatures. It has proved possible from large amounts of thermal ageing data to establish so-called generic thermal (or temperature) indices, which provide a comparative indication of the thermal stability of polymers for the purposes of materials selection (*see* table 5.6).

TABLE 5.6
Underwriters Laboratories thermal index (°C)

ABS	60
PA, PP	65
PC, PETP, PBTP	75
EP	90
UF	100
UP, silicone rubber	105
PCTFE, FEP, PF, MF, PSU	150
PTFE, PES	180
PEEK	240

5.6 High Temperature Polymers

The search for polymers with outstanding thermal stability has been vigorous, for obvious engineering reasons. Specialised materials have been successfully developed which owe their heat resistance to features of the primary chain structure. From the data of table 5.5 it appears that a heat-resistant polymer should melt above 300 °C and have useful thermal (and preferably oxidation) resistance in that region. These are severe requirements since some method of fabrication or processing must also be possible (the polymer must be stable in the melt or soluble or capable of formation *in situ*).

Diamond, which may be regarded as a completely crosslinked carbon polymer based on the C—C single bond, is stable to well over 1000 °C. At higher temperatures it transforms spontaneously into graphite, a carbon polymer based on flat sheets built of 6-carbon (aromatic) rings. Graphite is thermally stable at temperatures well over 2000 °C. These facts suggest that polymeric carbon-based materials of good thermal stability are possible, although in their physical properties they may exhibit some of the properties of ceramics.

The examples of diamond and graphite suggest that crosslinking and structural rigidity bring thermal stability. Strengthening the main chain by incorporation of aromatic 6-carbon and similar rings, and the synthesis of polymers with ladder-structure chains have yielded useful heat-resistant materials. Some structures are given in table 5.7. The aromatic polyimides and polyamides (*aramids*) are now important commercial materials.

5.7 Photo-oxidation

Early failures in outdoor uses of plastics and rubbers quickly revealed the severe effect of ultraviolet light on many polymers in the presence of oxygen. Figure 5.6 shows the distribution of energy in the solar spectrum (diffuse daylight). The molar photon energy is given on the lower abscissa scale. The C—C bond energy is about 330 kJ/mol corresponding to a photon wavelength of 360 nm. This means that light of this wavelength or shorter is capable of breaking C—C bonds in polymer molecules by which it is absorbed. 290 nm may be taken as the shortest wavelength (highest energy) radiation present in the ultraviolet spectrum of the sun *at ground level*. Radiation of shorter wavelength is removed by molecular oxygen and ozone in the atmosphere.

The mechanisms of oxidative photodegradation are probably diverse, and it is only possible to indicate the kind of processes involved. In fact the direct

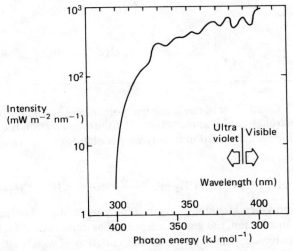

Figure 5.6 Intensity of solar radiation at the earth's surface: spectral distribution at blue and near-ultraviolet wavelengths. The lower scale shows the corresponding molar photon energy

TABLE 5.7

Structures of polymers with exceptional high temperature performance

		Upper service temperature (°C)
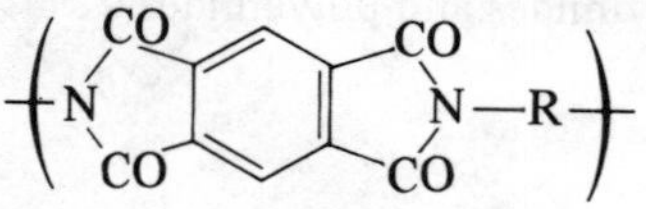	polyimide	300–350
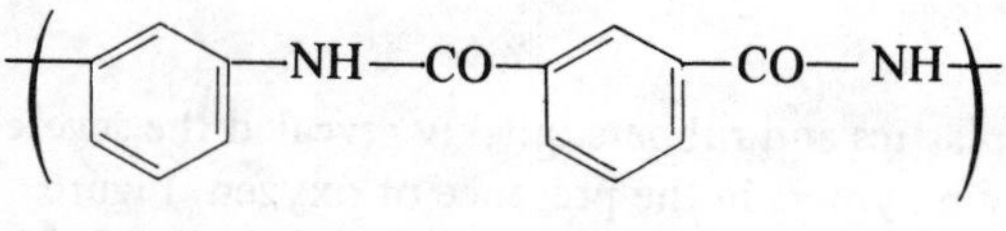	aromatic polyamide (aramid)	200–250
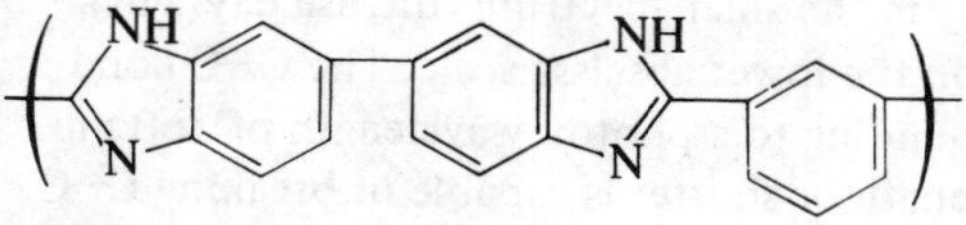	polybenzimidazole	250–300
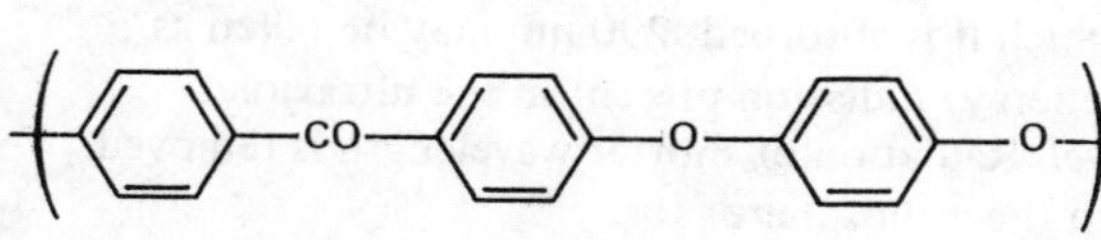	polyetheretherketone	240–260
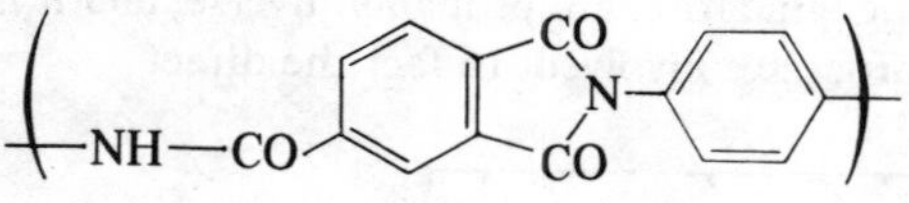	polyamide-imide	220–240
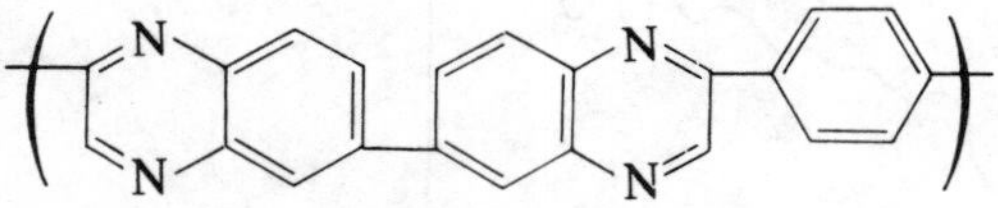	polyquinoxaline	400–450

breaking of single C—C and C—H bonds is relatively uncommon. The C=O bond absorbs more strongly and it appears that the following reaction is important in initiating photo-oxidation in polymers containing CO groups

$$R-\overset{\overset{\displaystyle O}{\|}}{C}-CH_2-CH_2-CH_2-R' \xrightarrow{\text{u.v.}} R-\overset{\overset{\displaystyle O}{\|}}{C}-CH_3 + CH_2=CHR'$$

Many engineering polymers including polyesters, polyurethanes and polyamides contain CO groups. In addition, CO groups are present as impurities in PE and other hydrocarbon polymers. The observation that radiation of wavelength longer than 340 nm does not produce photodegradation in PE is one important piece of evidence for this. Photo-oxidation leads generally to discoloration, surface cracking and deterioration of mechanical and electrical properties.

Controlled crosslinking by exposure to light is exploited in *photoresist* printing processes, used for example in making printed electronic circuits. The exposed areas of the photoresist polymer film are rendered insoluble and the image is developed by solvent etching. The use of additives to improve photo-oxidative durability is discussed in chapter 6.

5.8 Ageing and Weathering

No material is immutably stable. The processes of metallic corrosion have their counterparts in the science of polymers. Even the mildest of human environments may be extremely aggressive to a synthetic material. To compensate for this, the resistance to what are ordinarily regarded as highly corrosive substances may be excellent. Each application and each material must be considered carefully and objectively. Almost all polymers age at a significant rate unless steps are taken to stabilise them. In some cases the rate of deterioration of commercial formulations may set surprisingly short limits on the expected lifetime. Loss of appearance frequently occurs long before the component becomes functionally unserviceable. Quite reasonably the designer and the consumer seek reliable information on durability from materials suppliers. Polymer scientists and technologists have studied intensively the problems of predicting weathering performance.

The natural environment is enormously varied. Weathering is the result of exposure of polymers to conditions under which thermal oxidation and photo-oxidation may proceed simultaneously; and *in addition* the effects of water, abrasion and atmospheric pollution need to be evaluated. Microclimates may vary even over small distances, and features of the component such as its surface-to-volume ratio and the aspect of its surfaces may strongly influence weathering.

It is of course impossible to obtain information on long-term weathering performance quickly. Many possible methods of *accelerated testing* have been advocated, but have met with considerable criticism. Testing machines range from crude devices (in which, for example, test samples are simply exposed to high intensity visible/ultraviolet radiation) to elaborate environmental simulators. Inevitably, however, the processes of degradation are somewhat distorted by the intensification of the environmental conditions; the greater the acceleration the less the reliability. Despite this, accelerated test methods are widely employed, and for limited extrapolations may be useful.

5.9 Ionising radiation

Chemical change in polymeric materials may also be produced by *ionising radiations*. These effects are important in determining how polymers perform in radioactive environments, for example in nuclear reactors, or in space. In addition, the controlled use of radiation to produce beneficial changes in polymer properties during processing is of increasing importance.

Irradiation may be by gamma- or X-rays (high energy electromagnetic radiation, of wavelength shorter than ultraviolet), by high energy electrons, or by neutrons. The radioisotope ^{60}Co is an important gamma-ray source. The nuclear reactor produces both gamma-rays and neutrons. However, the electron accelerator is the most widely used source of ionising radiation for industrial applications in polymer processing. All such radiation interacts with polymer materials to cause ionisation by ejection of electrons from some molecules of the solid, directly or indirectly leaving these molecules in a highly energetic or *excited* state: $R \rightarrow R^*$.

The excited molecule may spread its energy among its neighbours as thermal vibration (dissipation as heat) or it may emit a photon. Sometimes the energy of excitation may become localised in a particular chemical bond, causing it to break. Radiation-induced scission of a primary chemical bond is called *radiolysis*. Such scission frequently causes the loss of side groups, as in the case of PE when hydrogen gas (H_2) is formed

$$\{CH_2-CH_2\} \xrightarrow{\text{radiation}} (CH_2-CH_2)^* \longrightarrow \{CH=CH\} + H_2$$

This reaction produces unsaturation in the polymer chain $(-C=C-)$. The cleavage of the CH bonds may occur on adjacent molecules with the important effect of crosslinking the chains

$$
\begin{array}{ll}
-CH_2- & \quad -\overset{\bullet}{C}H- \quad + H\cdot \\
-CH_2- & \quad -\overset{\bullet}{C}H- \quad + H\cdot
\end{array}
$$

$$
\begin{array}{l}
-CH- \\
\quad \quad \quad + H_2 \\
-CH-
\end{array}
$$

Crosslinking increases the molar mass of the polymer producing an increase in insoluble *gel* material, and may enhance thermal stability and alter mechanical properties markedly.

Alternatively radiolysis may involve a bond of the main chain; extensive chain scission leads to reduction in molar mass

$$
\begin{array}{lll}
-CH-CH_2- & \longrightarrow & -C=CH_2 + - \\
\mid & & \mid \\
CH_3 & & CH_3
\end{array}
$$

The sensitivity of a polymer to radiation effects is shown in the value of a parameter G, *the chemical yield*, the number of radiolytic reactions produced by the absorption of 100 eV of radiation. We may define G values for each of the processes shown above: $G(cl)$, $G(h)$, $G(cs)$ for crosslinking, for hydrogen evolution and for chain scission.

If $G(cl) \ll G(cs)$, the effect of radiation is to degrade the polymer. PMMA and PIB are notable cases. In PMMA the low molar mass compounds produced by breaking CH bonds or C—C bonds to side groups are trapped in the brittle solid, but are released when the solid is heated above T_g producing a foam. Irradiation produces extensive crosslinking in PE, POM, PP, natural rubber and some synthetic elastomers. Radiation crosslinking also occurs in PS, but this is the most resistant of the commodity polymers to radiation damage, as is shown by the G values of table 5.8. The main effect of irradiating PVC is to release HCl (cf. pyrolysis — *see* section 5.5) for which process G is about 13.

Ionising radiation is used commercially to produce crosslinked PE. Radiation penetrates deeply into the material, and the effects are relatively insensitive to temperature. This permits polymer chains to be crosslinked after the product is fabricated. Such radiation crosslinking increases thermal stability and reduces creep. Crosslinking probably occurs only in the amorphous regions of the polymer and the changes in elastic modulus which are observed depend on the degree of crystallinity. Crosslinked PE film stable up to 200 °C is manufactured.

Certain additives are capable of inhibiting these radiation effects, just as polymers may be stabilised against photo-oxidation and thermal oxidation (*see* chapter 6).

5.10 Combustion: Fire Properties of Polymeric Materials

All polymer materials are regarded as combustible, because at sufficiently high temperatures they undergo more or less complex reactions with oxygen, evolving

TABLE 5.8
Chemical yields for radiolytic reactions in common polymers

	$G(cl)$	$G(cs)$	$G(h)$
PE	2	0	4
PP	0.26	0.29	2.6
PS	0.04	0.01	0.02
PMMA	0	1.2	—
PIB	0	2.8	1.5
Nylon 66	0.7	2.4	0.5
POM	6.5	11.0	1.7
Natural rubber	3.5	0.5	0.6

heat and sometimes producing flame. The range of behaviour observed in polymers is once again very wide, and the assessment of fire properties extremely complicated. There is space here only to outline the main aspects of this subject.

Ignition and Combustion of Solid Materials

In contrast to the slow processes of thermal oxidation involved in ageing, combustion is understood to be a rapid process of oxidation which takes place above a critical temperature at which *ignition* occurs. In the case of most solid combustible substances, it is clear that the external heat source which brings about ignition serves to decompose the surface layers of the solid, producing combustible vapours (figure 5.7). In the case of polymers this process of decomposition is essentially the pyrolysis (perhaps assisted by oxygen) described in section 5.5. When the volatile pyrolysis products come into contact with air, an oxidation reaction occurs, with the liberation of heat. At a

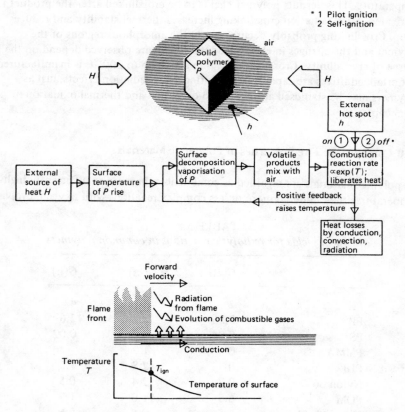

Figure 5.7 Mechanisms of ignition and flame spread in polymer materials

particular temperature (the ignition temperature) the rate at which heat is
released is sufficient to balance heat losses to the surroundings, and a stable
combustion process is established in the region where pyrolysis products and
atmospheric oxygen can interdiffuse. Usually this reaction zone is somewhat
luminous and we speak of it as a flame. Frequently, as in the case of PMMA or
PE rods burning in air, pyrolysis completely volatilises the material, and no
residue is left. In other cases, such as PVC, many crosslinked resins and wood, a
carbonaceous residue (char) remains. Such materials burn in two well-defined
stages. The initial diffusion flame near the surface is succeeded by a slower
combustion reaction within the porous char itself which occurs truly at the
interior surface, and the rate of which is controlled by oxygen diffusion into the
material.

It is clear even from this simple description that the combustion behaviour of
a polymer material results from the interplay of pyrolysis, gas phase oxidation
and flame gas dynamics, all acting in a strongly non-isothermal setting. It is not
surprising that test methods designed to assess fire properties are both numerous
and *ad hoc*. Such tests have the virtue of simplicity, but their value is frequently
limited by their arbitrariness. It is helpful for the technologist to understand the
test procedures fully to avoid reading too much meaning into test results.

In recent years the concept of the *limiting oxygen index* (as measured by the
Fenimore—Martin test) has been widely used in comparing the fire properties
of polymers. The apparatus is shown in figure 5.8(a). The specimen (in the form
of a vertical rod or panel) is ignited in an upward-flowing oxygen—nitrogen
mixture, and a stable candle-like flame is established, burning downwards. The
O_2/N_2 ratio of the gas mixture in the chimney is slowly reduced until the
candle-like flame becomes unstable and is extinguished. The minimum oxygen
fraction which can support steady combustion is termed the 'limiting oxygen
index' of the material. The parameter can be reproducibly measured and the
test procedure is simple. It must nevertheless be emphasised that the LOI
(unlike, for example, density or heat capacity) is not an inherent material
property but describes only the comparative behaviour of materials in a particular
test.

LOI values of a number of polymers are given in table 5.9. Since air contains
21 per cent oxygen, polymers having LOIs greater than 0.21 are self-extinguishing
in air *under conditions similar to those of the test*. Under fire conditions, where
some convective and radiative heat transfer usually assists burning, a figure of
0.27 or higher may perhaps indicate a useful degree of flame retardancy (*see also*
chapter 6).

In severe fires, all polymeric materials oxidise and make a contribution to
the heat liberated. The total contribution is determined by the *heat of
combustion* $\Delta H_c^\ominus$ (calorific value) which is a thermodynamic property of the
material. $\Delta H_c^\ominus$ is the quantity of heat liberated in the complete combustion of
unit amount of material. $\Delta H_c^\ominus$ gives no information about the rate at which
heat is released, although this is often critical in assessing fire hazard. Thus, for

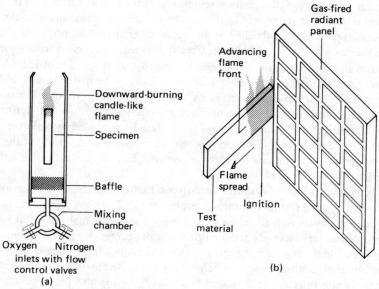

Figure 5.8 (a) Test apparatus for the measurement of the limiting oxygen index of polymers (Fenimore–Martin test). (b) Test configuration for the measurement of the surface rate of spread of flame of building materials

example, PE film and rod have the same $\Delta H_c^\ominus$, although film is more easily ignited and burns faster. Table 5.10 lists heats of combustion of a number of common polymers. $\Delta H_c^\ominus$ is controlled by molecular structure. Hydrocarbon polymers have the largest heats of combustion. Oxygen-containing polymers (wood, cellulosics and oxygen heterochains) release less heat on burning.

In many applications of polymer materials, notably in construction and textile uses, the rate at which flame spreads across a burning combustible surface is especially significant for safety. *Surface spread of flame rate*, although partially determined by inherent combustion properties of the material, is once again strongly dependent on circumstantial factors. Figure 5.7 shows that flame spreads by a succession of ignition processes, the flame zone invading non-burning areas as these reach sufficiently high temperatures. The rate at which heat is transferred to cooler parts of the material from burning parts depends on the efficiency of heat transfer processes. Generally, radiation from the flame, conduction through the solid and convection adjacent to the surface all make contributions to this heat transfer. Thus, flame luminosity, thermal diffusivity of the solid, and orientation of the surface (figure 5.7) are major factors controlling the surface spread of flame rate. The British Standard test for building materials classifies materials according to horizontal flame spread rate in the presence of

TABLE 5.9

Limiting oxygen indices of common polymer materials

POM	0.16
PMMA	0.17
PP	0.17
PE	0.17
PAN	0.18
PE + 50 wt% aluminium oxide	0.20
PE + 60 wt% aluminium oxide	0.30
PS	0.18
Cellulose	0.19
EP	0.20
EP + 50 wt% aluminium oxide	0.25
EP + 60 wt% aluminium oxide	0.41
PVF	0.23
PA-66	0.24
PC	0.27
Silicone rubber	0.30
PEEK	0.35
Polyimide	0.37
PPS	0.44
PVC	0.47
PVDC	0.60
PTFE	0.95

radiation from an external source — figure 5.8(b). A commercial 1.6 mm thick PVC sheet obtained a class 1 rating when bonded to 6 mm asbestos sheet, but the poorer class 2 rating both when bonded to 13 mm plywood sheet and when unsupported. Thinner PVC films on both steel and plasterboard can achieve the best rating, class 0. The fact that the assessment is based on horizontal not vertical spread emphasises the probably unavoidable arbitrariness of test procedures.

Flame Reactions

The chemistry of polymer flame reactions is not yet known in detail, although the reactions occurring in the flames of hydrocarbon fuels such as methane have been much studied. There is evidence that hydrocarbon polymers burn in much the same way as these fuels. The chemical changes occurring during combustion comprise complex sequences of gas-phase radical reactions, and undoubtedly all volatilising polymers burn by such reactions. (The burning of polymer chars,

TABLE 5.10

Heat (enthalpy) of combustion $\Delta H_c^{\ominus}$ *of selected polymers*

	$\Delta H_c^{\ominus}$ (kJ g^{-1})
POM	16.9
Cellulose: cotton linters	17.4
Cellulose: wood pulp	17.5
PVC	19.9
PETP	21.6
Polyurethane	23.9
PMMA	26.2
PA 66, PA 6	31.9
PC	30.8
PS	42.2
NR: pure gum vulcanisate	45.2
PP	46.0
PE	46.5

however, is a gas–solid *surface* reaction between oxygen and carbon.) Most flame-retardant additives act by interfering chemically with these flame reactions. These additives are discussed in chapter 6.

All radical reactions involve initiation, propagation and termination processes, and because of the large number of species present in polymer flames several parallel processes may occur. Somewhat schematically then we may describe the process of polymer combustion in the following way

$$polymer \xrightarrow{\text{heat}} decomposition\ products\ RH_2,\ R'H_2,\ etc.$$

where RH_2, $R'H_2$ denote various volatile fragments of the polymer chain. These fragments invariably contain H atoms which play an important part in the flame chemistry. For example, the flame reaction is initiated by the generation of a radical

$$RH_2 + O_2 \longrightarrow RH\cdot + \cdot HO_2$$

Once formed the radical can propagate the chain reaction; for example

$$RH\cdot + O_2 \longrightarrow RHO_2\cdot$$

$$RHO_2\cdot \longrightarrow RO + \cdot OH$$

$$\cdot OH + RH_2 \longrightarrow H_2O + RH\cdot$$

$$\overline{}$$

$$RH_2 + O_2 \longrightarrow RO + H_2O$$

The last line of this scheme shows the net chemical change which occurs as a result of the three preceding reactions (obtained simply by adding these reactions together). RO represents partially oxidised substances, which are further oxidised in combustion. An important intermediate of this kind is formaldehyde HCHO

$$HCHO + \cdot OH \longrightarrow CHO\cdot + H_2O$$

$$\cdot CHO + O_2 \longrightarrow CO + \cdot HO_2$$

Termination occurs by destruction of radicals such as $\cdot HO_2$ in the outer regions of the flame; for example

$$\cdot HO_2 + \cdot HO_2 \longrightarrow H_2O_2 + O_2$$

In fires, combustion is frequently incomplete, and volatile compounds and smoke are generated in greater or less amounts. The solid particles and liquid droplets present in smokes are produced by the condensation of organic substances (particularly acetylenes) in the cooler zones of the flame. Materials with high smoke yields are clearly hazardous in fires. In addition there is evidence that volatile toxic substances are produced when certain polymer materials burn. Carbon monoxide is always produced when carbon-based materials (including such natural substances as wood and wool as well as synthetic polymers) burn with a limited supply of air. Nitrogen-containing polymers (PUR, PA, PAN) may produce hydrogen cyanide under these conditions and chloropolymers may produce phosgene. However, the quantities evolved depend on the circumstances of the fire, and the hazards from toxic fire gases are extremely variable.

5.11 Biological Attack

Synthetic polymer materials generally show a remarkable resistance to attack by microbial organisms. Because of their comparative softness they are liable to attack by boring animals. Of common synthetic polymers only plasticised PVC is subject to microbial attack, and it is now clear that this results from the presence of the plasticiser rather than the polymeric PVC itself. Certain materials of animal and vegetable origin such as casein, cellulose and surface coating alkyds containing natural oils are attacked by bacteria or fungi.

5.12 Biocompatibility: Polymers in Medical Engineering

Polymers for use in surgery and biomedical engineering must be capable of performing, often for long periods of time, in contact with living tissue and body fluids. This is a highly specialised environment of great chemical and biochemical complexity, and is an interesting case study in polymer engineering. The principal medical uses of polymers are

(1) Structural materials in hip, knee, elbow and other reconstructions and replacements; lenses.
(2) Dental materials.
(3) Devices (including tubing) for transport of blood and other fluids, both inside and outside the body; and separation membranes (for example, for dialysis).
(4) Tissue adhesives.
(5) Sutures.

In all these applications, materials must of course be demonstrably non-toxic. Clearly, not only the polymer but also its possible degradation products, any residual monomer and all additives should be free of harmful effects. This requirement has not always been met. For example, it is known that methyl 2-cyanoacrylate (*see* section 6.7), an adhesive capable of bonding to wet tissue surfaces, produces a toxic reaction by yielding formaldehyde as it degrades. Other acrylates have proved satisfactory.

Polymer materials for artificial hip joints and similar uses must be not only biochemically compatible but also mechanically satisfactory. Friction and wear are the most critical aspects of mechanical performance and the most difficult properties to satisfy. The tribological operation of a natural human joint is highly evolved and efficient.

Even granted chemical inertness, it is well established that simple surface roughness can promote blood clotting. There is some evidence that other physical characteristics of the surface, such as ionic charge and wettability, also affect the *thrombogenicity* of polymer materials. The most successful methods of overcoming undesirable blood clotting involve binding a thin layer of the anti-coagulant mucopolysaccharide heparin to the polymer surface.

Figure 5.9 shows, as an example of precision medical engineering, a *heart-assist* device which makes use of a variety of polymer materials.

Suggestions for Reading

Solubility, Solution Properties and Permeability

ASTM D2857, *Standard Test Method for Dilute Solution Viscosity of Polymers* (American Society for Testing and Materials, Philadelphia, Pa, 1977).
Bixler, H. J. and Sweeting, O. J., 'Barrier properties of polymer films', in O. J. Sweeting (Ed.), *The Science and Technology of Polymer Films*, vol. 2, ch. 1 (Interscience, New York, 1971).
Burrell, H. and Immergut, B., 'Solubility parameter values', in J. Brandrup and E. H. Immergut (Eds), *Polymer Handbook*, ch. IV–15, 2nd edn (Wiley, New York, 1975).

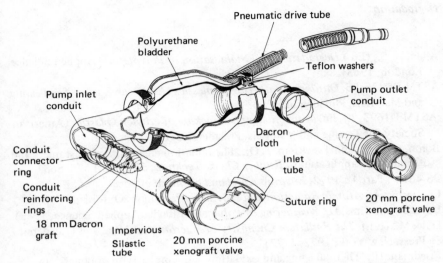

Figure 5.9 Polymeric materials in medical engineering: a left ventricle heart-assist device incorporating several polymers (by permission of Thermo Electron Corporation). The xenografts are pig aortic valves, processed with glutaraldehyde to crosslink the tissues. Teflon (PTFE) and Dacron (PETP fibre) are trade names of E.I. DuPont de Nemours Co. Inc.; Silastic (silicone rubber) is a trade name of Dow Corning Co. Inc.

Combellick, W. A., 'Barrier polymers', in *Encyclopaedia of Polymer Science and Engineering*, vol. 2, pp. 176–192 (Wiley, New York, 1985).

Comyn, J. (Ed.), *Polymer Permeability* (Elsevier, London, 1985).

Hennessy, B. J., Mead, J. A. and Stening, T. C., *The Permeability of Plastics Films* (Plastics Institute, London, 1966).

van Krevelen, D. W., *Properties of Polymers: Correlations with Chemical Structure* (Elsevier, Amsterdam, 1972).

Nemphos, S. P., Salame, M. and Steingiser, S., 'Barrier polymers', in *Encyclopaedia of Polymer Science and Technology*, suppl. vol. 1, pp. 65–95 (Wiley, New York, 1976).

Ninnemann, K. W., 'Measurement of physical properties of flexible films', in O. J. Sweeting (Ed.), *The Science and Technology of Polymer Films*, vol. 1, ch. 13 (Interscience, New York, 1968).

Stannett, V. and Yasuda, H., 'Permeability', in R. A. V. Raff and K. W. Doak (Eds), *Crystalline Olefin Polymers*, pt II, ch. 4 (Interscience, New York, 1964).

Tanaka, T., 'Gels', *Scientific American*, **244** (1981) 124–138.

Yasuda, H., Clark, H. G. and Stannett, V., 'Permeability', in *Encyclopaedia of Polymer Science and Technology*, vol. 9, pp. 794–807 (Wiley, New York, 1968).

Degradation

Allen, N. S., (Ed.) *Degradation and Stabilisation of Polyolefins* (Applied Science, London, 1983).

ASTM D1435-85, *Outdoor weathering of plastics* (American Society for Testing and Materials, Philadelphia, Pa.).

ASTM D1693-70, *Environmental stress-cracking of ethylene plastics* (American Society for Testing and Materials, Philadelphia, Pa.).

Böhm, G. G. A. and Tveekrem, J. O., 'The radiation chemistry of elastomers and its industrial applications', *Rubber Chem. Technol.*, **55** (1982) 575–668.

BS 4618 : Part 4 : 1972, *Environmental and chemical effects.*

Charlesby, A., *Atomic Radiation and Polymers* (Pergamon, Oxford, 1960).

Davis, A. and Sims, D., *Weathering of Polymers* (Elsevier Applied Science, 1983).

Dole, M., (Ed.), *The Radiation Chemistry of Macromolecules*, 2 vols (Academic Press, New York, 1972, 1973).

Friedman, H., 'Thermal aging and oxidation with emphasis on polymers', in I. M. Kolthoff, P. J. Elving and F. H. Stross (Eds), *Treatise on Analytical Chemistry: Part III, Analytical Chemistry in Industry* (Wiley, New York, 1976).

Grassie, N. and Scott, G., *Polymer Degradation and Stabilisation* (Cambridge University Press, 1985).

Hawkins, W. L., (Ed.), *Polymer Stabilization* (Wiley, New York, 1972).

Howard, J. B., 'Fracture: long-term phenomena', in *Encyclopaedia of Polymer Science and Technology*, vol. 7, pp. 261–291 (Wiley, New York, 1967) (environmental stress cracking).

Institute of Metals, 'Environmental cracking of polymers', in *Mechanics of Environment Sensitive Cracking of Materials*, pp. 213–259 (London, 1977).

Jellinek, H. H. G. (Ed.), *Aspects of Degradation and Stabilization of Polymers* (Elsevier, Amsterdam, 1978).

Kelen, T., *Polymer Degradation* (van Nostrand Reinhold, New York, 1982).

Keller, R. W., 'Oxidation and ozonation of rubber', *Rubber Chem. Tech.*, **58**, (1985) 637–652.

Küster, E., 'Biological degradation of synthetic polymers', *J. Appl. Polymer Sci.: Appl. Polymer Symp.*, **35** (1979) 395–404.

Lanza, V. L., 'Irradiation – properties changes', in R. A. V. Raff and K. W. Doak (Eds), *Crystalline Olefin Polymers* pt II, ch. 7 (Interscience, New York, 1964).

Owen, E. D. (Ed.), *Degradation and Stabilisation of PVC* (Applied Science, London, 1984).

Rånby, B. and Rabek, J. F., *Photodegradation, Photo-oxidation and Photo-stabilization of Polymers* (Wiley, London, 1975).

Reich, L. and Stivala, S. S., *Elements of Polymer Degradation* (McGraw-Hill, New York, 1971).

Rosato, D. V. and Schwartz, R. T., *Environmental Effects on Polymeric Materials*, 2 vols (Interscience, New York, 1968).

Schnabel, W., *Polymer Degradation: Principles and Practical Applications* (Hanser, Munich, 1981).

Winslow, F. H., 'Environmental degradation', in J. M. Schultz (Ed.), *Treatise on Materials Science and Technology*, vol. 10, pt B, ch. 6 (Academic Press, New York, 1977).

Thermal Stability

Cassidy, P. E., *Thermally Stable Polymers* (Dekker, New York, 1980).

Critchley, J. P., Knight, G. J. and Wright, J. J., *Heat-Resistant Polymers* (Plenum, New York, 1983).

UL 746B, *Standard for polymeric materials — long term property evaluation*, 2nd edn (Underwriters Laboratories, Melville, NY, 1979).

Fire Properties

Aseeva, R. M. and Zaikov, G. E., 'The flammability of polymeric materials' in *Advances in Polymer Science*, vol. 70, *Key Polymers — Properties and Performance*, pp. 171–229 (Springer-Verlag, Berlin, 1985).

ASTM D2863-77, *Minimun oxygen concentration to support candle-like combustion of plastics (oxygen index)* (American Society for Testing and Materials, Philadelphia, Pa.)

Cullis, C. F. and Hirschler, M. M., *The Combustion of Organic Polymers* (Clarendon, Oxford, 1981).

Fabris, H. J. and Sommer, J. G., 'Flammability of elastomeric materials', *Rubber Chem. Tech.*, **50** (1977) 523–569.

Fenimore, C. P. and Martin, F. J., 'Burning of polymers', in *The Mechanisms of Pyrolysis, Oxidation and Burning of Organic Materials*, National Bureau of Standards Special Publication no. 357, pp. 159–170 (Washington DC, 1972).

Gann, R. G., Dipert, R. A. and Drews, M. J., 'Flammability' in *Encyclopaedia of Polymer Science and Engineering*, vol. 7, 2nd edn, pp. 154–210 (Wiley, New York, 1987).

Hilado, C. J., *Flammability Handbook for Plastics*, 3rd edn, (Technomic Publishing Co., Westport, Conn., 1982).

Kuryla, W. C. and Papa, A. J., (Eds), *Flame Retardancy of Polymeric Materials*, 5 vols. (Dekker, New York, 1973–79).

Lewin, M., Atlas, S. M. and Pearce, E. M. (Eds) *Flame-Retardant Polymeric Materials*, vol. 1 (Plenum, New York, 1975); vol. 2 (Plenum, New York, 1978); vol. 3 (Plenum, New York, 1982).

Stuetz, D. E., Diedwardo, A. H., Zitomer, F. and Barnes, B. P., 'Polymer combustion', *J. Polymer Sci.*, pt A, **13** (1975) 585–621.

Troitzsch, J., *International Plastics Flammability Handbook* (Hanser, Munich, 1983).

Medical Engineering

Chiellini, E., Giusti, P., Migliaresi, C. and Nicolais, L., *Polymers in Medicine II: Biomedical and Pharmaceutical Applications* (Plenum, New York, 1986).

Frazza, D. J., 'Sutures', in *Encyclopaedia of Polymer Science and Technology*, suppl. 1, pp. 587–597 (Wiley, New York, 1976).

Gebelein, C. G. (Ed.) *Advances in Biomedical Polymers* (Plenum, New York, 1987).

Halpern, B. D. and Karo, W., 'Medical Applications', in *Encyclopaedia of Polymer Science and Technology*, suppl. 2, pp. 368–402 (Wiley, New York, 1977).

Kronenthal, R. L., Oser, Z. and Martin, E., (Eds), 'Polymers in Medicine and Surgery', in *Polymer Science and Technology*, vol. 8 (Plenum, London, 1975).

Williams, D., (Ed.), *Biocompatibility of Implant Materials* (Sector Publishing, London, 1976).

6

Polymer Materials and their Technology

Engineering design, whether of buildings or of microelectronics, is very much a matter of exploiting the properties of materials for practical ends. The engineer expresses his skill as a designer at every stage through the *selection of materials*. Early in the evolution of a design solution he considers alternative materials in broad terms: metal or non-metal, ceramic or polymer, ferrous or non-ferrous metal. He makes his choice from general considerations of engineering properties, including cost and ease of processing and fabrication. Later he gives attention to his options within a selected class of materials, and he requires more precise comparative information on those properties which emerge as determinative: perhaps the corrosion resistance of certain alloy steels or the dielectric properties of a group of thermoplastics. Ultimately, the designer has to examine in detail the specifications of a number of *commercial* materials, including possibly the nominally identical products of competing manufacturers.

The earlier chapters of this book have dealt with the characteristics of polymers as a major class of engineering materials. The emphasis has been on the scientific basis of materials properties. The diversity of polymer materials has become apparent and the reader will now be familiar with the principal kinds of polymers. In this last chapter we turn to the technology of polymers — the production of commercial materials with refined properties, in a variety of forms for engineering use, and the methods of processing and fabrication.

6.1 Engineering Thermoplastics

The principal types of commercially available thermoplastics are listed in table 6.1. The classification is based on the primary chemical structure, but each polymer type is marketed in numerous grades and modifications, so that the number of individual materials runs into several thousands. Some of the most important modifications are also included in table 6.1. Historically, the line of development of commercial thermoplastics started with the materials based on cellulose. A crucial early innovation was the Hyatt brothers' use of camphor as a plasticiser for cellulose nitrate CN. Subsequently, as the major commodity homopolymers

TABLE 6.1

The principal commercially available engineering thermoplastics

Thermoplastics	Modifications
High density polyethylene HDPE	Homopolymer and copolymer,
Low density polyethylene LDPE	linear low density forms,
	ultra-high molecular weight
Ethylene/vinyl acetate copolymer EVA	
Ionomer	
Polypropylene homopolymer PP	High impact forms
Polymethylpentene PMP	
Polystyrene PS	Homopolymer
	High impact (toughened) HIPS
Acrylonitrile–butadiene–styrene ABS	General-purpose, high impact
Styrene–acrylonitrile copolymer SAN	
Poly(vinyl butyral) PVB	
Poly(vinyl chloride) PVC	Rigid (unplasticised vinyl homopolymer)
	Rigid, rubber modified
	Flexible compounds (plasticised vinyl)
	Dispersion resins (latex, suspension, emulsion)
	Chlorinated PVC, CPVC
	Copolymers
Polytetrafluorethylene PTFE	Granular forms
	Unfilled homopolymer
	Filled forms
	Dispersions
Poly(vinyl fluoride) PVF	
Polychlorotrifluorethylene PCTFE	
Poly(vinylidene fluoride) PVDF	
Polyamides PA	Nylon 6 (PA 6), nylon 610 (PA 610), nylon 66 (PA 66), copolymers
	Unfilled and filled grades
	Aramids

TABLE 6.1 continued

Thermoplastics	Modifications
Poly(methyl methacrylate) PMMA	
Polyoxymethylene POM	Acetal homopolymer
	Acetal copolymer
Polysulphone PSU, Polyethersulphone	
Polyetheretherketone PEEK	
Polyester–urethane	
Polyether–urethane	
Poly(phenylene oxide) PPO	PPO/PS and other blends
Poly(phenylene sulphide) PPS	
Polycarbonate PC	
Polyimide PI, Polyetherimide PEI	
Poly(ethylene terephthalate) PETP	
Poly(butylene terephthalate) PBTP	
Polyarylate	
Cellulose acetate CA	
Cellulose acetate butyrate CAB	
Ethyl cellulose	

(Thermoplastic elastomers: *see* section 6.5)

PVC, PS and PE were introduced, the manufacturers overcame deficiencies in their engineering properties by the use of additives. Plasticisers were developed which greatly altered the mechanical behaviour of PVC; mineral fillers were found to improve weathering and increase rigidity, and stabilisers to prevent thermal decomposition during melt processing. PE was stabilised against photodegradation for outdoor use. Property-modifying additives are now almost invariably incorporated in commercial thermoplastics. We discuss the function of additives and the *compounding* of polymers and additives more fully in the following section.

From the early cellulose-derived materials right through to the present the development of new polymers with novel primary chain structures has continued. However, commercial materials of new structure are infrequent arrivals on the polymer scene (as table 1.2 showed). In parallel, established materials are repeatedly improved and modified by innovations in synthesis and in compounding. One particularly fruitful and important theme in polymer modification has been the development of the *hybrid polymer*. The first important example of a hybrid material was high impact polystyrene HIPS. The early exploitation of straight PS was bedevilled by its brittleness and attempts were made in the 1940s to improve its performance by blending PS with rubbers. Some success was achieved with simple two-phase physical mixtures, but in 1952

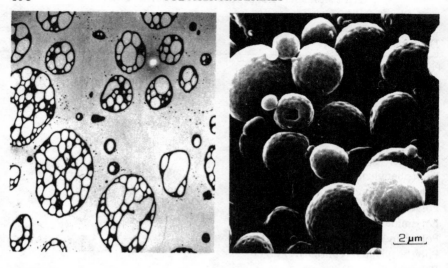

Figure 6.1 Morphology of high impact polystyrene HIPS. (a) Transmission electron micrograph, showing dispersed elastomer particles (~ 2 μm diameter) with PS occlusions. (b) Scanning electron micrograph of elastomer particles after dissolving PS matrix. (Courtesy of Dow Chemical Company)

the Dow Chemical Co. introduced a two-phase HIPS in which a graft copolymer is formed at the interface between the PS matrix and dispersed SBR or BR elastomer particles (figures 6.1 and 6.2). A little later the first of another major type of rubber-toughened materials, the ABS thermoplastics, was introduced commercially. These polymers also have a two-phase morphology and consist of elastomer particles dispersed in a styrene–acrylonitrile SAN copolymer matrix. These trends towards hybrid polymer materials led in the 1960s and 1970s also to the emergence of polymer alloys, a name applied to rubber-toughened materials in which the matrix is a mixture of polymer types. Among established polymer alloys are ABS/PVC, ABS/PC, ABS/polysulphone, PMMA/PVC and PPO/PS (modified PPO). Thus rubber toughening, copolymerisation, blending and alloying have emerged as methods of polymer modification of great importance and a current area of rapid growth. The characteristics of the various types of hybrid polymer are summarised in table 6.2.

We can illustrate the role of chain structure, chain length, additives and hybridising in polymer modification with the example of polypropylene PP and its derivative thermoplastics. Figure 6.3 shows the relations between the commercially important members of the PP family. Atactic PP produced by the high pressure polymerisation of propene proved to have no commercial value as a thermoplastic although it has limited use as an adhesive. On the other hand, Ziegler–Natta catalysts promote stereospecific synthesis and isotactic homo-

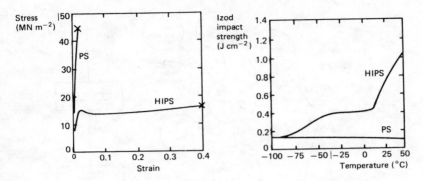

Figure 6.2   PS and HIPS: comparison of yield and impact behaviour

polymer PP became a major commodity material within a decade of its introduction. Chain length (molar mass) controls melt viscosity: commercial grades of differing melt viscosity are suited to various processing and fabrication techniques. Additives modify other properties such as fire performance, and thus further commercial grades exist for specific end-uses. In particular, for food and medical application additive toxicity is a primary consideration. Glass, talc and other filled grades are produced to meet the demand for a more rigid or cheaper bulked material. The brittleness of PP which is particularly marked at low

TABLE 6.2
Types of hybrid polymer materials

Physical blend	Compatible blends	1-phase
	Incompatible blends	2-phase
Copolymer	Random or alternating	1-phase
	Graft	Usually 2-phase
	Block	Often 2-phase, often small domain size
Alloy		Usually 2-phase graft, with physical blend matrix of two compatible polymers

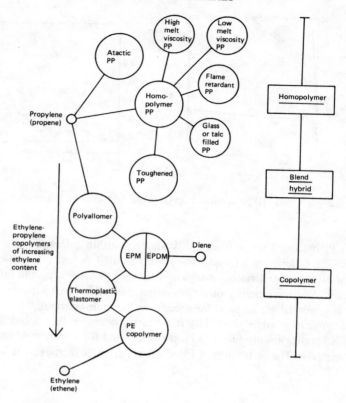

Figure 6.3   Members of the polypropylene family of thermoplastics and elastomers

temperatures and is a serious disadvantage for certain applications may be greatly reduced by copolymerisation. Random copolymers of propylene and ethylene (EPM) or terpolymers of propylene, ethylene and a diene (EPDM) are elastomers, and of major importance in their own right (*see* section 6.5). Toughened or high impact grades of PP may be produced by the mechanical blending of PP with EPM or EPDM. More recently new toughened PP materials (known as polyallomers) have been introduced which are formed by the direct copolymerisation of propylene and ethylene (2–10 per cent) in such a way that block copolymer structures are formed. Figure 6.4(a) shows the marked improvement in impact performance which results from incorporating small amounts of ethylene to form a block copolymer.

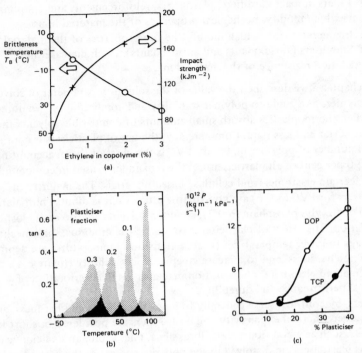

Figure 6.4   (a) The effect of ethylene copolymerisation on the impact strength and impact brittleness temperature of PP. (b), (c) The effect of plasticisation on the properties of PVC: (b) the anelastic spectrum; (c) the water vapour permeability (dioctyl phthalate and tricresyl phosphate plasticisers)

6.2 Compounding of Thermoplastics

The main types of additives incorporated in polymer materials are as follows:

Antimicrobials and biocides	Impact modifiers
Antistats	Lubricants
Blowing agents	Plasticisers
Colourants	Reinforcements
Fillers	Slip and antiblock agents
Flame retardants	Stabilisers

While it is clear that almost every material property can be modified, the additives of the greatest general importance fall into four groups.

(1) Lubricants and heat stabilisers which assist processing.

(2) Fillers, impact modifiers, plasticisers, reinforcements and coupling agents which modify mechanical properties of the material in use.
(3) Flame retardants which modify the fire properties of the material.
(4) Stabilisers (antioxidants and antiozonants) which increase the degradation resistance of the material in use.

In chapter 5 we described the value of the solubility parameter of solvents and polymers as a guide to polymer solubility and compatibility. We noted that when thermoplastics absorb small amounts of compatible solvents they become softer and less rigid. Commercially, the use of plasticisers is confined to a small number of polymer materials. PVC is the chief of these, accounting for about 90 per cent of all plasticiser use; the remainder is used in compounding PVAC, certain elastomers and cellulose-based materials. The majority of plasticisers for PVC (δ 9.6) are non-volatile esters such as dibutyl phthalate (δ 9.3) or trixylyl phosphate (δ 9.7); common solvents are too easily lost by evaporation. The effect of plasticiser is to reduce the modulus, shift mechanical loss peaks to lower temperatures (that is, transition temperatures are depressed), to decrease hardness, and to increase creep and permeability (figure 6.4). As we have remarked elsewhere, plasticisation may also entail deterioration of electrical performance and durability.

Fillers, by contrast, serve generally to increase modulus and hardness, and to reduce creep, thermal expansivity and cost. Fillers are powdered or short-fibre materials, such as wood flour, glass fibre, silica, talc, kaolin and calcium carbonate. Whereas plasticisers are dissolved in the polymer and act at the molecular level (reducing interchain forces for example), the filler particles form a distinct phase in a mechanical mixture. They exert their influence on the properties of the compounded solid through composite action with the polymer matrix.

The flame retardant and stabiliser additives are substances chosen to interfere with the chemical reactions underlying combustion, oxidation and other modes of chemical degradation. The variety of compounds used for these purposes is enormous. We can do no more than indicate by way of illustration (figure 6.5) the mode of action of two typical additives.

Synergistic Bromine—Antimony Flame Retardants for PP

OH· and other radicals which participate in the combustion flame reactions (*see* reactions 5.6 and 5.7, p. 136) may be removed by inhibitors such as hydrogen bromide, HBr. HBr is formed *in situ* from the decomposition of a bromine-containing organic compound (such as hexabromocyclodecane HBCD) incorporated in the polymer as a flame-retardant additive. The effectiveness of such flame retardants (as measured by their effect in raising the limiting oxygen index) is frequently enhanced by the presence of antimony oxide Sb_2O_3 (an example of synergism in the action of additives) — figure 6.5(a).

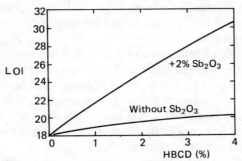

Figure 6.5(a)   Mode of action of flame retardant additive HBCD of the flame poison type

Under flame conditions the following reactions occur

$$HBCD + Sb_2O_3 \longrightarrow SbBr_3 \text{ antimony bromide}$$

$$SbBr_3 \longrightarrow Br\cdot \text{ radical}$$

$$Br\cdot + \text{polymer pyrolysis products} \longrightarrow HBr$$

$$OH\cdot + HBr \longrightarrow H_2O + Br\cdot$$

Thermal Antioxidants

These are additives which stabilise polymers against thermal oxidation (*see* section 5.4). As shown in figure 6.5(b) an important group of these anti-oxidants act by removing $RO_2 \cdot$ radicals, thus retarding the chain reaction given on page 146.

OH

R″   R″

$\xrightarrow{RO_2^{\cdot}}$ RO_2H +

R′

Stable product

Phenol antioxidant additive

O•

R″   R″

$\xrightarrow{RO_2}$

R′

Relatively unreactive radical intermediate

O

R″   R″

R′   OOR

Stable product

Figure 6.5(b)   Mode of action of a thermal antioxidant additive

The list of additives is completed by a number of minor entries. Antimicrobials are biocides admixed to prevent fungal and other types of biological attack.

These substances are necessarily somewhat toxic and their inclusion may not be permitted for certain end-uses. However, they are frequently incorporated into surface coating formulations.

Antistats (commonly glycerol esters) act to reduce triboelectric charging by lowering resistivity. Blowing agents are used in forming cellular (foamed) plastics and are discussed in section 6.6 below. Reinforcements are considered in sections 6.9 and 6.10 under the headings of *fibres* and *composites*. Colourants embrace pigments (insoluble, finely divided solids) and soluble dyes, selected for colour stability, inertness and lack of toxicity. Pigments may act to some extent as fillers; their optical properties are discussed in section 6.8.

6.3 Processing and Fabrication of Thermoplastics

Processing and fabrication describe the conversion of materials from stock form (bar, rod, tube, pellet, sheet and so on) to a more or less complicated artefact. Whatever the material the shaping process must involve one or more variants of cutting, joining and moulding. Wood is almost invariably worked by cutting and joining operations. Techniques for shaping metals are very diverse; methods of cutting and joining are highly developed, but casting, sintering, pressing and rolling processes are also widely used. Moulding techniques predominate in the technologies of glass and ceramics. Polymer materials have proved especially amenable to a variety of extrusion and moulding techniques. In particular the following principal processing and fabricating operations for thermoplastics now enable products and components of complex shape to be mass-produced on a very large scale

Extrusion
Injection moulding
Blow moulding
Rotational moulding
Calendering
Thermoforming (sheet moulding)
Casting
Sintering
Machining
Cold-forming
Welding
Cementing

The most important processing and fabricating techniques for thermoplastics exploit their generally low melting temperatures and shape the materials from the melt. *Extrusion* and *injection moulding* are the most widely used processes, and they are illustrated in figure 6.6(a, b). The screw extruder accepts raw thermoplastics material in pellet form and carries it through the extruder barrel; the material is heated by contact with the heated barrel surface and also by the mechanical action of the screw, and melts. The melt is compressed by the taper of the screw and is ultimately extruded through the shaped die to form tube, sheet, rod or perhaps an extrusion of more complicated profile. The screw extruder works continuously and the extruded product is taken off as it emerges from the die for reeling or cutting into lengths. It is important that the melt viscosity be sufficiently high to prevent collapse or uncontrolled deformation of the extrudate when it leaves the die, and there may be water or air sprays at the outlet for rapid cooling. High melt viscosities can be obtained by using materials of high molar mass. The rate of cooling of an extrusion may determine the degree of crystallinity in a crystalline polymer, and hence affect mechanical and other properties.

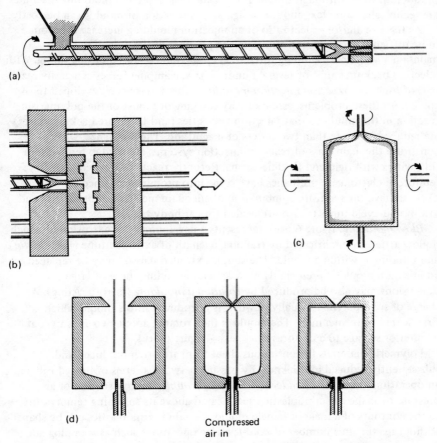

Figure 6.6   Operating principles of machine tools for (a) extrusion, (b) injection moulding, (c) rotational moulding and (d) blow moulding of thermoplastics

Injection moulding describes a process in which polymer melt is forced into a mould, where it cools until solid. The mould then separates into two halves to allow the product to be ejected; subsequently the parts of the mould are clamped together once more, a further quantity of melted material is injected and the cycle repeated. The injection end of the machine is most commonly an archimedean screw (similar to that of a screw extruder) which can produce, once per cycle, a shot of molten polymer of predetermined size and then inject it into the mould by means of a reciprocating ram action. Low melt viscosity is desirable to obtain good flow within the mould cavities, but high injection temperatures mean longer cooling periods, longer cycle times and lower production rates. Mould and product design are influenced by consideration of polymer flow patterns, orientation and crystallisation effects. Injection moulding provides a particularly effective way of obtaining complex shapes in large production runs. Although simple in principle large injection moulding machines are technically complex, and the design and fabrication of moulds can be costly.

As the size and/or aspect ratio of an injection moulding increase it becomes more difficult to ensure uniformity in the polymer during injection and to maintain a sufficient clamping force to keep the mould closed during filling. With injection back pressures of several hundred bars, clamping forces can easily reach tens of tons. The *reaction injection moulding* process has been developed to overcome both these problems, essentially by carrying out most of the polymerising reaction in the mould so that injection viscosities (and therefore back pressures) are reduced by more than two orders of magnitude. The urethanes, which constitute the main class of reaction injection systems, are used extensively to fill cavities with rigid and flexible foams. Recently reaction-injected glass fibre reinforced elastomeric urethanes have come into prominence, since they seem likely to have the requisite combination of adequate mechanical properties and low mould-cycle times ($<$ 1 min) needed for car body panels.

Blow moulding, figure 6.6(d), represents a development of extrusion in which hollow articles are fabricated by trapping a length of extruded tube (the *parison*) and inflating it within a mould. The simple extruded parison may be replaced by an injection-moulded preform. Hollow articles including those of large dimensions may also be produced by *rotational moulding* (or rotocasting). A charge of solid polymer, usually powder, is introduced into a mould which is first heated to form a melt. The mould is then rotated about two axes to coat its interior surface to a uniform thickness − figure 6.6(c).

Polymeric materials in continuous sheet form are often produced and subsequently reduced in thickness by passing between a series of heated rollers, an operation known as *calendering. Thermoforming* employs suction or air pressure to shape a thermoplastic sheet heated above its softening temperature to the contours of a male or female mould. Certain thermoplastics can be shaped without heating in a number of *cold-forming* operations such as stamping and forging commonly applied to metals.

All these methods of fabrication are essentially moulding processes. Cutting techniques (embracing all the conventional machining operations: turning, drilling, grinding, milling and planing) can also be applied to thermoplastics, but they are used much less widely. In comparison with metalworking, difficulties arise from the low melting temperatures, low thermal conductivities and low moduli of polymer materials which in consequence are relatively easily deformed by the cutting tool. Cutting actions are less precise than on metals and heating at the tool surface must be avoided. Nevertheless with appropriate depth of cut, rate of feed and tool geometry conventional machine tools may often be used. (The relatively poorer performance of mechanical fastenings also militates against cutting and joining as a general method of fabrication for thermoplastics.) These techniques are most commonly applied to glassy polymers such as PMMA and to thermoplastic composites such as glass-filled PTFE, although softer materials such as PE and unfilled PTFE have excellent machinability.

Many thermoplastics may be satisfactorily cemented either to similar or dissimilar materials (adhesives are discussed in section 6.7). A number of inert and insoluble polymer materials such as PTFE, PCTFE and some other fluoropolymers, PE, PP and some other polyolefins are amenable to cementing only after vigorous surface treatment. For the lower melting thermoplastics welding provides an important alternative for joining parts of the same material, for example in fabricating pipework. The thermoplastics of higher melting temperature (which include PTFE of the common engineering polymers and also some of the specialised heat-tolerant materials such as the polyimides) cannot be satisfactorily fabricated by the principal extrusion and injection moulding processes. These materials are shaped by *sintering* powdered polymer in pressurised moulds, a process which causes the polymer particles to coalesce. Sintered products are normally somewhat more porous than those made from thermoplastics which are processed by way of the melt.

6.4 Thermosets

Thermoset materials are produced by the direct formation of network polymers from monomers, or by crosslinking linear prepolymers. Once formed the polymer cannot be returned to a plastic state by heating. For thermosets therefore the polymerisation (or at least its final *curing* stage) and the shaping process occur simultaneously. Long-established and relatively simple processing methods for thermosets include casting and compression and transfer moulding. More recently, automatic injection moulding of thermosets has been successfully developed and is now widely used.

The principal thermosets are the phenolics, the amino resins (UF and MF), the epoxies, the unsaturated polyesters (including the surface coating alkyds described in section 6.8) and the crosslinked polyurethanes. Of less importance are the crosslinked silicones, the furan resins and the allyl resins, notably allyl diglycol carbonate. The chemistry of the thermoset resins is generally more complicated than that of the thermoplastics. All are formed by the reaction of at least two types of monomer, usually in a sequence of polymerisation and crosslinking stages. Unlike the thermoplastics processor, the fabricator working with thermoset materials has to carry out the final curing reactions.

We have already described in outline in section 1.14 how thermosets of the unsaturated polyester and epoxy type are formed. In both these cases the first stage of polymerisation is the construction in a condensation reaction of a short linear chain (the relative molecular mass generally lies between 500 and 5000, corresponding to 2–12 monomer units). The uncured resins are stable and are made available to the user as liquids of various viscosities, as pastes or, in the case of the longer chain epoxies, as solids with melting temperatures up to 150 °C.

The final curing of the resins is achieved by combining the UP or EP resin immediately before moulding with hardeners or curing agents which crosslink the polymer chains through reactive groups. The C=C double bonds of the UP chains are coupled through styrene crosslinks by the action of free radical initiators. The epoxies crosslink through the terminal $>C-C<$ groups by reaction
$$O$$
with amines or anhydrides. Different curing agents require different curing conditions and curing times. The various base resins in combination with selected curing agents produce thermosets having somewhat different properties. Therefore an epoxy or polyester system comprising a base resin and curing agent may be tailored to a particular application, for which certain electrical, mechanical or thermal properties and also cost are optimised as appropriate. For example, fire performance of polyesters is improved by replacing phthalic acid by tetrachlorophthalic acid. Impact strength is increased by incorporating carbon chain acids such as adipic acid $HOOC(CH_2)_8COOH$ in the UP polymer.

The crosslinking of both UP and EP resins occurs via addition reactions and no by-product is generated. By contrast, condensation reactions occur at each stage in the thermosetting of PF and amino resins and water is generated as a reaction product. Figure 6.7 shows how phenolic resins are produced and cured.

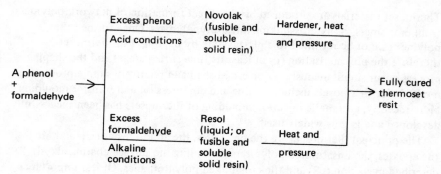

Figure 6.7   Novolak and resol routes to phenolic thermosets

The *novolak* formed in the two-stage process is a useful partially polymerised but still fusible intermediate. This can be combined with a filler such as wood flour and a crosslinking agent or hardener (normally hexamethylene tetramine) to form a moulding powder, which is finally fully cured only under the action of heat and pressure in the mould. Phenolic laminates are usually made from *resols*, in what is described as a one-stage process, since the addition of a hardener is not necessary. The condensation polymerisation proceeds essentially continuously from raw materials to the fully cured *resit*, although intermediate stages in the development of the network polymer are recognised and the polymerisation may be interrupted at the resol stage.

6.5 Elastomers

The defining characteristic of elastomers as a class is their high elasticity — the capacity for very large rapidly recoverable deformation. As we have seen in chapter 3 such behaviour appears in lightly crosslinked amorphous polymers above their glass transition temperatures. Elastomeric mechanical behaviour was first fully realised in natural rubber NR vulcanised by reaction with sulphur (Goodyear and Hancock, about 1840). Rubber rapidly proved an immensely useful material and soon became indispensable to the industrial nations. Germany and the United States in particular sought urgently to develop synthetic substitutes to secure their resources of this vital material and from the 1920s onwards a succession of *synthetic rubbers* were developed. Finally in 1954 a stereoregular *cis*-1,4-polyisoprene essentially identical to natural rubber was synthesised, using Ziegler–Natta and other catalysts. This synthetic polyisoprene rubber IR was subsequently put into successful commercial production.

All these synthetic materials were closely modelled on natural rubber, being linear chain polydienes and random copolymers, with C=C double-bond unsaturation in the chain to permit irreversible crosslinking, essentially as in natural rubber itself (table 6.3). The residual unsaturation of the unvulcanised chain is achieved by using a diene monomer with two C=C double bonds. Only half this unsaturation disappears during polymerisation, leaving one C=C bond per diene monomer unit. For example, polybutadiene

More recently (notably with the introduction of polyurethane elastomers (AU and EU) in the early 1950s) other materials have appeared, which while possessing high elasticity differ considerably from the rubbers in other properties and in structure. Some of these elastomers are close structural relations of thermoplastics and thermosets. Some, such as the Y class *thermoplastic elastomers*, based on block copolymers, represent major advances in polymer materials. Taken together these trends indicate a marked diversification in elastomer technology. Table 6.4 lists elastomer materials of current commercial importance (*see also* figure 1.1 for world consumption data).

Despite the competition of highly developed synthetic materials, natural rubber has retained a leading place among commodity and engineering elastomers, and production increases year by year. The growth of the rubber industry from the seeds of the wild rubber tree of Brazil is a remarkable achievement of agriculture and applied chemistry. Rubber cultivation consists in the production of a stereoregular hydrocarbon polymer as a *crop*, by biosynthesis, using solar energy, carbon dioxide, water and the elaborate metabolic reactions of plant tissue: a model perhaps of future materials biotechnologies. The interesting techniques which were evolved in the nineteenth and early twentieth centuries to convert raw rubber latex to useful commercial rubbers are outlined in table 6.5. These operations provide the basis also for the technology of synthetic rubbers.

Natural rubber latex is a dispersion of rubber particles ($0.1-1.6$ μm diameter) in water. Unvulcanised raw rubber obtained by coagulating and drying the rubber latex has a very high molar mass (with average chain lengths of the order of 100 000 C atoms) and before compounding the rubber is *masticated*. Mastication, which involves mechanically working the rubber for some minutes, reduces the chain length by a factor of about 10, and converts the raw gum to a plastic mass. In this state, the rubber can be compounded with additives, including most importantly vulcanising agents, stabilisers and reinforcing fillers. Sulphur, the original vulcanising agent for natural rubber, remains pre-eminent, but its effectiveness has been enhanced and brought under control by the use of accelerators and activators.

The stabilisers are substances which inhibit the oxidative ageing and ozone attack to which natural rubber is subject. They act by intercepting active free radicals and breaking the sequence of degradation reactions (*see* sections 5.4 and 6.2). Amines and phenols are used. The reinforcing fillers, of which finely divided carbon powders (*carbon blacks*) are by far the most important, serve to improve mechanical end-use properties such as modulus and hardness, abrasion and tear resistance (figure 6.8). The mechanism of reinforcement remains somewhat obscure; it is clear from measurements of the viscosity and solubility of unvulcanised mixes that the rubber adheres strongly to the carbon surface, and generally the reinforcing effects of carbon blacks increase with decreasing particle size. Furnace blacks with particle diameters around $0.02-0.06$ μm are most commonly used. Carbon blacks are also highly efficient in absorbing ultraviolet light and some also significantly increase electrical conductivity, reducing triboelectric charging and acting as antistats. Inert fillers such as pigments and extenders are frequently included in the compounded rubber mix, to which may also be added process and extender oils — *see* table 6.6.

The vulcanising agent reacts with the unsaturated polyisoprene chains of NR forming crosslinks and tying the molecules together chemically at certain points

TABLE 6.3
Primary chain structure of the major synthetic rubbers

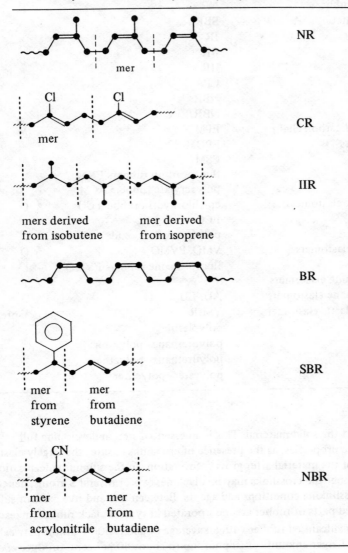

TABLE 6.4
Commercial elastomers

Type		ASTM class
Natural rubber	NR	R
and synthetic	SBR	
polydienes	IR	
	BR	
	IIR	
	CR	
	NBR	
	NBR/PVC	
Saturated carbon chain	EPM	M
elastomers	EPDM	
	CSM	
	fluorocarbon rubbers FKM	
	polyacrylate rubbers ACM	
Polyether elastomers	epichlorhydrin rubbers CO	O
	ECO	
	poly(propylene oxide) rubbers GPO	
Silicone elastomers	VMQ, PVMQ	Q
	fluorosilicone rubbers FVMQ	
Polysulphide elastomers		T
Polyurethane elastomers	AU, EU	U
Thermoplastic elastomers	YSBR	Y-prefix
	polyolefin	
	polyurethane–polyester	
	polyurethane–polyether	
	polyester–polyether	

throughout the solid material. This is an essential step in developing full elastomeric properties, as the presence of crosslinks ensures the largely elastic recovery of the material after gross deformation. In the sulphur vulcanisation of natural rubber the crosslinks may involve one, two or several S atoms, depending on the crosslinking conditions and agents. Between one and five parts of sulphur per hundred parts of rubber are incorporated in typical black rubber mixes: this produces a vulcanised rubber with an average of about 500 C atoms between crosslinks. Larger amounts of sulphur, up to about 40 per cent, produce *ebonites* or hard rubbers, which are highly crosslinked, rigid nonelastomers.

The vulcanisation of NR is invariably assisted by heat, so that processing normally consists of mixing, forming and heat-curing stages. In the forming of

TABLE 6.5
Compounding and processing operations in rubber technology

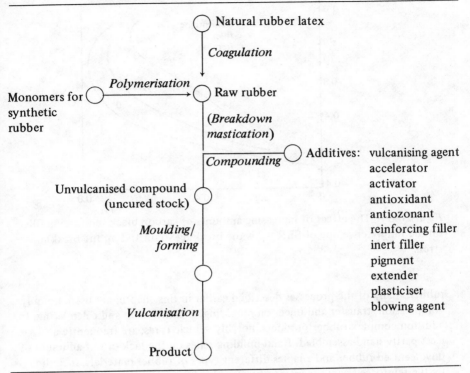

TABLE 6.6
An example of a compounded rubber for tyre use

Component	Parts by weight	Function
Raw natural rubber	100	Elastomer
Stearic acid	2.5	Accelerator–activator
Zinc oxide	3.5	Accelerator–activator
Amines	6.0	Antioxidant/antiozonant
Mineral oil	5.0	Plasticiser
Wax	2.0	Processing aid
Carbon black	50	Reinforcing filler
Sulphur	2.5	Vulcanising agent
Amide	0.5	Accelerator

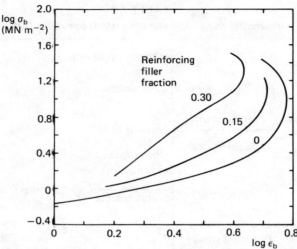

Figure 6.8 The effect of increasing amounts of carbon black reinforcing filler on the failure envelope of SBR. ϵ_b is the breaking strain and σ_b the breaking stress

rubber, many of the processes described earlier in this chapter are used, notably compression, transfer and injection moulding, and extrusion and calendering. In addition complex rubber products, notably vehicle tyres, are frequently at least partly hand-assembled. Hand-building relies on the tack or self-adhesion of unvulcanised rubber and enables different types of rubber materials, including textile-reinforced components, to be assembled into a composite structure. Finally whatever the forming process heat-curing, either within the mould or in a separate autoclave, completes the sequence of processing operations.

The technology of the polydiene and polyolefin synthetic rubbers (R and M class) broadly follows the pattern established for NR. Vulcanising by sulphur is commonly applied to SBR, NBR, IR, BR, IIR and EPDM. Zinc and magnesium oxides are the normal vulcanising agents for CR: crosslinking probably occurs through an —O— link. The saturated chain elastomers such as EPM and the fluoroelastomers FKM can be crosslinked by various organic substances; for example, peroxides remove hydrogen atoms from occasional chain branch points in EPM to produce crosslinks.

Carbon blacks are also widely used as reinforcing fillers in these elastomers. Their use is especially important in those synthetic rubbers which, unlike NR, do not crystallise at high strains (chapter 3). SBR, BR and NBR are of this kind; in consequence their unfilled gum vulcanisates show little tensile strength (SBR about 2 MPa). Reinforcement with suitable carbon blacks can improve tensile strengths as much as ten-fold to values comparable with those of NR mixes (20 MPa).

All the elastomers so far described are carbon-chain polymers of quite simple primary structure, either homopolymer or copolymer, produced by free radical chain reactions. In some contrast the polyurethane elastomers (class U) form a major group of heterochain polymer materials of considerable structural diversity. The structural basis of the AU and EU elastomers is outlined in a simple form in table 6.7. The range of structure in these materials arises from the way in which several different monomers are incorporated in successive step reactions. The characteristic polyurethane grouping —NH—CO—O— appears in the polymerised structure linking either linear polyester or polyether segments derived from hydroxyl-terminated prepolymers by reaction with a diisocyanate. These segments form the major part of the ultimate elastomer. The prepolymer usually has a relative molecular mass of 1000—2000, corresponding to a chain length of about 100 atoms in the backbone. Some solid polyurethane elastomers (millable types) are processed by conventional milling and vulcanising operations; others are formed by casting directly from a liquid reaction mixture, and do not pass through the raw gum stage. The solid polyurethane elastomers are strong and have exceptional abrasion and tear resistance. In comparison with the polydiene and polyolefin rubbers and in consequence of their distinctly different primary chemical structure they have excellent resistance to oxidation, including ozone attack. The greater polarity of the heterochain structure also confers valuable resistance to aliphatic hydrocarbon solvents and oils.

Another major class of heterochain elastomers, the silicone SI rubbers (ASTM designation VMQ, FVMQ and others), are notable for their exceptionally wide working temperature range, extending as high as 200 °C for many purposes. They are largely unaffected by atmospheric degradation, and the fluorosilicone elastomers FVMQ show exceptional chemical inertness.

Unvulcanised natural rubber, although somewhat soft, manifests definite elastomeric properties, despite its lack of chemical crosslinks. This is attributed to physical entanglements of the exceptionally long polymer chains which it contains (*see* chapter 3). These entanglements are less permanent than the chemical ties present in the vulcanisate, and are slowly released by thermal

TABLE 6.7

Polyester—urethane (AU) and polyether—urethane (EU) elastomers

---

Urethane group —NH—CO—O— is formed by reaction of an *isocyanate* and an *hydroxy compound*:

$$R—NCO + HO—R' \longrightarrow R—NH—CO—O—R'$$
isocyanate   urethane

Thus a diisocyanate and a diol combine to yield a urethane polymer: Ⓐ

$$OCN—R—NCO + HO—R'—OH \longrightarrow \text{⟮}CO—NH—R—NH—CO—O—R'—O\text{⟯}$$
linear polyurethane

TABLE 6.7 continued

In synthesising polyurethane elastomers the hydroxy compounds used are either *polyether* or *polyester* intermediates with —OH groups at the chain ends (OH-terminated polymers):

HO — ~~~~~~ — OH

polyether (B)
or (linear or lightly branched)
polyester (C)

e.g.	polytetrahydrofuran	$HO[(CH_2)_4O]_n$—H a linear polyether
	poly(ethylene adipate)	$HO[CO(CH_2)_4CO-O-CH_2-CH_2-O]_nH$ a linear polyester
		(relative molecular mass typically 1000–2000)

Chain extension is achieved by linking polyether or polyester intermediates through urethane groups by reaction with a diisocyanate:

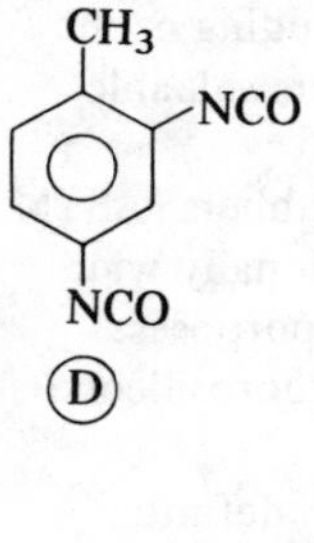

(D) + HO — ~~~~~~ — OH polyether (B) or polyester (C) ⟶ – NCO terminated polyether *or* polyester prepolymer

(D)—(B)—[(D)—(B)]$_n$—(D)

or

(D)—(C)—[(D)—(C)]$_n$—(D)

Three principal types of urethane elastomers are then produced:

Millable elastomers (processed by conventional rubber technology)	Uncured elastomers are long-chain polymers made by chain extension of polyether or polyester intermediates using a diisocyanate; curing is through NCO— groups using glycols
Cast elastomers	Liquid or easily melted short-chain —NCO terminated prepolymers; chain extension and curing by glycols
Thermoplastic elastomers (thermoplastics technology)	Polyether–polyurethane ((B)–(A)) or polyester–polyurethane ((C)–(A)) block copolymers

motion, so that raw rubber exhibits marked creep and set in long-term tests. However, in short timescale measurements (such as rebound) it shows elastomeric behaviour.

If physical entanglements in an amorphous polymer above T_g lack permanence, do there exist other physical means of tying molecular chains together to produce elastomers? The successful development since the mid-1960s of a series of commercial *thermoplastic elastomers* shows decisively that this is so. The materials of this class are now available are based on styrene—butadiene, polyether, polyester and polyurethane block copolymers. Ethylene—propylene thermo-plastic elastomers are probably of the same type. The styrene—butadiene materials have been most fully studied and characterised. They consist of A—B—A styrene—butadiene—styrene triblock chains, containing about 30 per cent styrene. The central 1,4-polybutadiene (soft block) segment typically has a relative molecular mass of 30 000—100 000 and the glassy polystyrene (hard block) segments are considerably shorter (10 000— 30 000). The solid has a two-phase morphology, showing spherical domains of about 0.01—0.03 μm diameter formed by the PS blocks within a matrix of polybutadiene, as shown in figure 6.9. (Different morphological structures including rods and sheets are found at other styrene/butadiene ratios.) These hard block domains act as physical crosslinks in forming the elastomeric network. In addition, they behave as a well-dispersed, fine-particle reinforcing filler in promoting high tensile strength and modulus. The effectiveness of these crosslinks diminishes rapidly above the T_g of polystyrene (~ 100 °C). At higher temperatures the materials become thermoplastic, and as a result can be processed by the standard processing techniques for thermoplastics (*see* section 6.3) — a most important property. There is evidence that the separate microphase domains persist in the melt, but they are of course fluid.

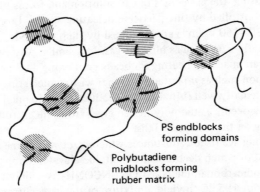

PS endblocks
forming domains

Polybutadiene
midblocks forming
rubber matrix

Figure 6.9   Schematic microstructure of an ABA type thermoplastic elastomer based on hard polystyrene endblocks and soft polybutadiene midblocks

6.6 Cellular Polymers

Cellular forms of many plastics and elastomers have been developed, largely for their thermal and mechanical properties. The technology has been applied most extensively to polyurethanes, polystyrene, poly(vinyl chloride), polyethylene, polypropylene, the phenolics, urea–formaldehyde resins, silicones, natural rubber and SBR synthetic rubber. Both *rigid* and *flexible* cellular materials are manufactured.

Most of the many processes for producing cellular polymer materials belong to one of three groups.

(1) Those in which voids are formed by dispersing air in a polymer emulsion or partially polymerised liquid by mechanical agitation.

(2) Those in which gas bubbles are formed within a liquid form of the polymer, generally a melt or dispersion, by heating or by reduction of pressure (the use of a *physical blowing agent*).

(3) Those in which gas bubbles are generated within the polymer liquid by a chemical reaction (the use of a *chemical blowing agent*).

Methods of the first group are used to produce foam from NR and SBR latices and also from UF prepolymer. Mechanical foaming generally produces materials with an *open-cell* structure, which have high permeabilities to liquids and gases. Methods of the second and third groups which employ blowing agents provide finer control of cell structure and can be used to form open-cell or *closed-cell* foams according to conditions. Closed-cell materials have much higher elastic moduli in compression than similar open-cell foams as a result of the presence of trapped air. Such materials also have excellent buoyancy and barrier properties. The bulk of cellular polymers are manufactured by processes which make use of blowing agents. Physical blowing agents are gases or volatile liquids which may be dissolved in the polymer under pressure; on releasing the pressure or on heating, the gas comes out of solution and forms bubbles, the size of which can be controlled by fine-particle nucleating agents. Low density polyethylene foams ($30-160$ kg/m^3) are produced by such methods, commonly using dichlorotetrafluoroethane as blowing gas. In a similar way cellular polystyrene materials are made by heating PS beads containing 4–8 per cent of a volatile hydrocarbon such as *n*-pentane together with a nucleating agent. These processes are particularly well suited to the production of foamed stock by extrusion; foaming occurs as the polymer emerges from the extruder nozzle and the pressure acting on the extrudate falls.

Chemical blowing agents are substances which react with some component of the polymer liquid or which decompose at suitable elevated temperatures to produce a gas. Azodicarbonamide $H_2NCON=NCONH_2$ is widely used; it decomposes at about 195 °C to yield a mixture of the gases nitrogen and carbon monoxide, with a little carbon dioxide. The total gas yield of 1 g of azodicarbonamide is about 230 cm^3 at STP – figure 6.10(a). Phenolic resins are foamed by the decomposition of the blowing agent on contact with acid acting as poly-merisation catalyst. Thus carbonates release carbon dioxide and metallic aluminium yields hydrogen.

The production of very low density foams is limited by the problem of instability and foam collapse during processing. Some thermoplastics foams can be stabilised by crosslinking, a technique which is used to make PE foams with chemical blowing agents. The crosslinking is achieved either by exposing the polymer to ionising radiation or by including a reactive crosslinking agent such as dicumyl peroxide as an additive — figure 6.10(b).

In certain cases the addition of a blowing agent may be unnecessary, for the polymerisation or curing reaction may itself form a gaseous by-product. The most important examples are the release of carbon dioxide by reaction of isocyanates with hydroxyl groups in the formation of polyurethane foams; and the release of hydrogen in room-temperature-foaming silicone resins and rubbers.

We have already noted (section 3.20) the very low thermal conductivity which can be achieved in low density cellular materials, which are therefore important insulants. The structural uses of rigid foams have expanded with injection moulding techniques and methods of forming integral-skin foamed products. As figure 6.11 illustrates, cellular forms can exhibit much greater flexural rigidity than the corresponding solid plastics, weight for weight. Rigid foams are thus

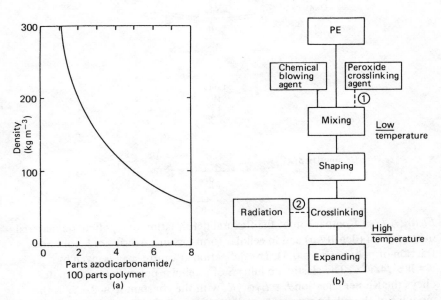

Figure 6.10   (a) The effectiveness of azodicarbonamide blowing agent in producing foams from low density polyethylene LDPE. (b) Flow diagram of route to cellular polyethylene crosslinked by (1) peroxide or (2) ionising radiation

increasingly used in sandwich panel composites with steel, aluminium or solid polymer facing sheets. Foamed elastomers (natural and synthetic rubbers and polyurethanes) are highly compressible, resilient materials used widely for gasketing, cushioning and sound and vibration damping.

6.7 Adhesives

Synthetic polymers have brought a great transformation in the technology of adhesives. About four decades ago, adhesives were few: bituminous cements and mastics; rubber cements; gelatine-based glues manufactured from hide, hoof and bone; casein glues from milk protein; and polysaccharide glues produced from vegetable starch. All these types had shortcomings. Bituminous cements have a long history of use in building and civil engineering, but are

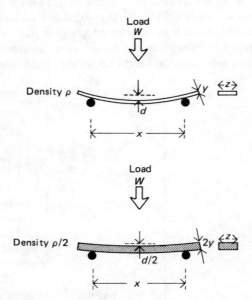

Figure 6.11 Comparison of the flexural deformation of a polymeric material in solid form (density ρ) and in cellular form (assume density $\rho/2$, volume fraction of polymer $\phi_1 = 0.5$). The deflection at centre span is given by $d = Wx^3/4zy^3E$. The elastic modulus E of a cellular polymer approximately obeys the Kerner equation $E = E_1\phi_1/K$, with the constant $K \approx 2$. E_1 is the modulus of the solid polymer. Note that the tensile strength of the cellular material is generally considerably less than the non-cellular material. This weakness is overcome in sandwich-panel construction or integral-skin foams by presence of stronger material on the tension face

subject to creep and soften seriously when warm. The animal and vegetable glues are vulnerable to fungal attack and either dissolve or swell on contact with water. Today however there is a great range of commercial adhesives (over 100 types and several thousand products are listed in the *Adhesives Handbook*) and most of the principal thermoset and thermoplastic synthetic polymers and elastomers find some application in adhesive formulations. In table 6.8 therefore we include only those polymers which have a major use as adhesives. Adhesive formulations are of five general types;

(i) Solutions of thermoplastics (including unvulcanised rubbers) which bond by loss of solvent.

(ii) Dispersions (emulsions) of thermoplastics in water or organic liquids which bond by loss of the liquid phase.

(iii) Thermoplastics without solvents (hot-melt adhesives).

(iv) Polymeric compositions which react chemically after joint assembly to form a crosslinked polymer (thermoset).

(v) Monomers which polymerise *in situ*.

Adhesives of types iv and v are converted by chemical reaction within the assembled joint. Chemical bonding between the adhesive and the adherend has sometimes been assumed; but adhesive bonding does not generally entail primary chemical bond formation across the interface. The reaction usually involves only components of the adhesive, which is cold-cured or heat-cured, with or without pressure. Thermoset adhesives which evolve water during curing (such as phenolic and amino resin types) usually require the application of pressure. Adhesives which are sufficiently polar to show large dielectric losses at high frequency may be successfully heat-cured by exposing the assembled joint to a radiofrequency electric field. Dielectric heating is widely used in bonding wood with PF, MF and UF adhesives. Cyanoacrylate adhesives are based on alkyl 2-cyanoacrylate monomers such as

$$CH_2{=}CCOOCH_3 \quad \text{methyl 2-cyanoacrylate}$$

with a CN group on the central carbon

which polymerise extremely rapidly in the presence of traces of moisture normally present on adherend surfaces. As noted in table 6.8, polymer hybrid (*two-polymer*) adhesives are of considerable importance. Elastomers are added to thermoset formulations to improve flexibility and impact resistance: examples are EP–polysulphide, PF–CR and PF–NBR. PF–EP blends have exceptional performance at temperatures up to 200 °C.

As the new adhesives have emerged the performance (especially the strength and permanence) which can be achieved in adhesive-bonded structures has greatly improved. The rational design of bonded joints has become a matter of concern to design engineers; there has also been parallel progress in the underlying science of adhesion, so that the factors which control adhesive action are becoming clear. It is essential that the adhesive should wet the adherend surface

TABLE 6.8
Polymers for adhesives use

Natural rubber NR
Butyl rubber IIR and polyisobutylene PIB
Acrylonitrile—butadiene copolymer NBR
Styrene—butadiene copolymer SBR
Thermoplastic elastomer YSBR
Polychloroprene CR
Polysulphide
Silicone polymers, Q-type elastomers SI

Phenol—formaldehyde PF
Resorcinol—formaldehyde
Amino resins UF, MF
Epoxy resin EP
Polyurethane resins
Unsaturated polyester resins including alkyds
Aromatic polyimides

Poly(vinyl acetate) PVAC and poly(vinyl alcohol) PVAL
Vinyl acetate—ethylene copolymer EVA
Polyethylene PE
Polyamide PA
Poly(vinyl butyral) PVB
Acrylic acid and acrylate ester homopolymers and copolymers
Poly(alkyl cyanoacrylate)
Cellulose derivatives and starch
Proteins (gelatine, casein, soybean, blood)

Note: In many adhesive formulations two or more polymer types are blended or co-reacted.

during the assembly of a bonded joint; and this depends on the surface energy of the adherend and the surface tension of the fluid adhesive. The ideal strength of the joint is controlled by the microstructure of the interfacial zone, and the van der Waals' forces acting within it. Adherend surfaces may be porous or impermeable, and vary widely in roughness. The adhesive/adherend line of contact may be microscopically sharp or may be diffuse, through mutual solubility. In practice joint strengths are reduced by the presence of flaws, which may lie either within the adhesive or sometimes within a weak surface layer in the adherend.

6.8 Surface Coatings

The range of application of polymeric materials as coatings is vast: almost all the major types of thermoplastics, thermosets and elastomers find use as coatings in some form. In many of these applications, such as PVC extrusion coating of wire or unsaturated polyester UP finishing of wood, the polymer materials and methods of processing do not differ much from those we have discussed earlier in this chapter. There is one major class of polymeric surface coating materials, the paints and varnishes, which are sufficiently distinct to warrant separate treatment. We devote this section to a brief survey of the science and technology of paints and related materials, and the polymers on which they are based.

Paint manufacture is a very old-established industrial activity. In its early phases its raw materials were exclusively natural in origin — natural oils, natural resins and mineral pigments, from which paints and varnishes were produced by blending and milling. The industry has been transformed in recent decades by the availability of synthetic polymers, synthetic pigments and a wide range of property-modifying additives. Materials of natural origin still find a place, but a diminishing one, in modern paint formulations.

A paint is a suspension of solid pigment particles in a liquid phase (the *vehicle*) which when applied to a surface dries to form a solid film. In the dried paint film the pigment is dispersed in a continuous matrix (or *binder*) which is invariably polymeric. Thus the vehicle is the liquid precursor of the binder. A varnish is simply an unpigmented paint. The function of the surface coating may be protective or decorative, or both; the polymeric binder is selected for a variety of qualities, including strength, hardness and abrasion resistance over a working range of temperature; weathering performance and resistance to chemical attack; and optical properties such as refractive index and transparency. Equally important are the properties of the liquid vehicle, which must be able to form stable dispersions with pigment, be of suitable viscosity and show a satisfactory rate of drying.

The binder materials which find use in paint formulations are very diverse, and it is useful to classify them according to the nature of the film formation process, or *drying mechanism* (*see* table 6.9). The simplest of these is evaporative drying, class A. The vehicle may be a solution of a polymeric solid in an organic solvent, and in such cases the wet film dries simply by loss of solvent to the air. The so-called nitrocellulose lacquers (based on solutions of cellulose nitrate CN in a solvent such as butyl acetate), chlorinated rubber dissolved in hydrocarbon solvents, and acrylic lacquers based on plasticised PMMA and copolymers, are all of this class. Also extremely important members are the *emulsion* or *latex* paints, in which the vehicle is a two-phase mixture of finely dispersed particles of polymer in water. Styrene—butadiene copolymers, poly(vinyl acetate) PVAC and copolymers, and poly(methyl methacrylate) copolymers are the three main kinds of polymer emulsion binder used. As the water evaporates from the wet film the stability of the dispersion is destroyed and the particles of polymer

TABLE 6.9
Classification of surface coatings by drying mechanism

A Evaporative drying (film formation without chemical reaction)

Solution types
    Cellulose nitrate CN (nitrocellulose)
    Cellulose acetate butyrate CAB
    Ethyl cellulose
    Poly(vinyl chloride-*co*-vinyl acetate)
    Poly(methyl methacrylate) PMMA and other poly(acrylic ester)s
    Chlorinated rubber
    Bituminous coatings

Dispersion types
    Emulsion binders based on
    Poly(styrene-*co*-butadiene)
    Poly(vinyl acetate) and copolymers
    Poly(methyl methacrylate) and copolymers
    PVC organosols

B Convertible coatings (film formation by chemical reaction)

B1 Air-drying types (oxygen reaction)
    Fatty drying oils
    Oleoresinous coatings
    Air-drying alkyds

B2 Reaction coatings (cold-curing or thermosetting by reaction of vehicle
    components)
    Epoxy EP
    Unsaturated polyester UP
    Polyurethane
    Urea– and melamine–formaldehyde UF, MF with and without alkyds

---

coalesce on contact to form a continuous film into which the pigment is bound.
For this to occur satisfactorily the glass transition temperature T_g of the polymer
should lie some degrees below the ambient temperature. This may be achieved
by temporarily plasticising the polymer (and thus depressing T_g) by incorporating
a small amount of fugitive plasticiser in the formulation. This plasticiser is slowly
lost by evaporation in the final stages of drying after the film is formed, causing
T_g to rise and improving the hardness of the paint surface.

Emulsion paints are a major development because they are water-based, and
avoid the toxicity and flammability hazards of paints which contain volatile

organic liquids. However, dispersion coatings are not exclusively water-based. PVC (the homopolymer of which has low solubilities in common solvents) is widely applied as a coating in the form of an *organosol*, a dispersion of PVC particles (~ 0.1 μm diameter) in organic liquids such as ketones. The T_g of PVC is about 80 °C so that heat is applied to bring about the formation of a continuous film.

In surface coatings in which film formation is by evaporative drying, the binder must be a thermoplastic since it must either be soluble or sinter as the emulsion film dries. Amorphous polymers are, as a rule, more soluble than crystalline and form transparent films, so that amorphous thermoplastics are commonly found in surface coating binders. Straight PS is not a satisfactory binder because it is brittle and lacks chemical resistance to solvents.

All other drying mechanisms involve chemical reaction, although simultaneous solvent loss may also occur. We can divide the chemically drying paints into two subclasses, B1 and B2. In the first, the film-forming components in the vehicle react with the oxygen in the air, or less commonly with water vapour. This subclass contains the traditional paint formulations based on natural *fatty oils* (such as linseed and soya and mostly of vegetable origin) and natural resins, together with their more recent modifications. In all these paints the reaction of the oils with atmospheric oxygen to produce crosslinked network polymers lies at the heart of the film-forming process.

Table 6.10 summarises in simple outline the broad characteristics of these complex materials and the paints made from them. The fatty oils themselves are all mixtures of triglycerides of C_{18} fatty acids, i.e. esters formed between the alcohol glycerol and the long-chain acids which are denoted by HOOC ∼∼∼∼∼∼∼∼ . There are many such acids (a few structures only are shown): they differ largely in the number and position of the C=C bonds, which of course play a decisive part in crosslinking the oil molecules during drying. As can be seen, linseed oil (one of the leading paint oils) contains about 50 per cent of linolenic acid which has three C=C bonds, 15 per cent of linoleic acid (with two C=C bonds) and 20 per cent of oleic acid (with one C=C bond). On exposure to air, linseed oil absorbs oxygen which initiates a series of free radical reactions, producing a network polymer crosslinked through C—O—O—C, C—O—C and C—C primary chemical bonds. Film formation in a paint or varnish may be speeded by partially polymerising the oil before incorporating it into the vehicle, and assisted by additives. In practice, oils are not generally used on their own as binders because the resulting film lacks water resistance and strength. Traditional *oleoresinous* paints and varnishes are based on mixtures of drying oils with various resins, either natural (such as rosin or shellac) or synthetic (notably phenol–formaldehyde PF novolaks and resols). In some cases the resins react chemically with the oil during the manufacture of the vehicle; in other cases the resins simply dissolve in the oil. In both, oxidative drying of the fatty oil component leads to the formation of a resin-containing film, which shows improvements in properties such as drying speed, hardness or water resistance.

TABLE 6.10
Natural drying oils

Natural fatty oils are mixtures of compounds of structure (I):

$$CH_2O\text{—}CO\text{——}\qquad CH_2OH$$
$$CHO\text{—}CO\text{——}\qquad CHOH \qquad \text{————}COOH$$
$$CH_2O\text{—}CO\text{——}\qquad CH_2OH$$
$$\text{(I)}\qquad\qquad\qquad \text{(II)}\qquad\qquad \text{(III)}$$

These compounds are esters of glycerol (II) and C_{18} fatty acids represented by (III); the most important of these fatty acids and their distribution in three major natural oils are:

	Fatty acid composition (% by weight)		
	Linseed oil	Soya bean oil	Tung oil
Oleic acid $CH_3(CH_2)_7CH=CH(CH_2)_7COOH$	20	22	13
Linoleic acid ~~~~~~COOH	20	50	—
Linolenic acid ~~~~~~COOH	50	9	—
Elaeostearic acid ~~~~~~COOH	—	—	80

The natural oils react with atmospheric oxygen to form network polymers crosslinked through the C=C double bonds

Natural oils + resins $\longrightarrow$ oleoresinous paints and varnishes

Natural oils + glycerol + phthalic anhydride $\longrightarrow$ alkyd paints

(*see* table 6.11)

One chemical modification of the fatty drying oil binder deserves special mention: this is a class of branched chain polyesters, known universally in the coatings industry as *alkyds*, which have been outstandingly important in the history of paint development. Air-drying alkyds are formed (table 6.11) from fatty oils, additional glycerol and an organic acid or anhydride, such as phthalic anhydride. Great variation is possible in the structures of alkyds according to ingredients and synthesis conditions, but in essence esterification of the acid and glycerol provides a polyester backbone along which the fatty oil chains are attached. The subsequent crosslinking of these side chains through the reactive C=C double bonds produces the insoluble network polymer film.

The other subclass of paints in table 6.9 which dry by chemical reaction is B2; this includes all those coatings in which the solid film is formed by reaction between components of the vehicle itself. B2 paints are thus mixed immediately before they are applied to bring the reactive ingredients together, or else converted by heat, ultraviolet light or electron beam irradiation and are based essentially on thermoset formulations. Perhaps the most versatile are epoxy EP and acrylic non-aqueous dispersion coatings, but urea–formaldehyde UF and melamine–formaldehyde MF, unsaturated polyester UP and polyurethane types are also of major importance. The formulations are diverse and complicated, often involving copolymerisation with alkyds, and the details are outside the scope of this account. However, they supply a wide range of factory-applied, high performance finishes for innumerable manufactured products, including automobiles and furniture.

We have discussed in outline the polymer materials which are used as paint film formers. A complete paint formulation (such as that given in table 6.12 for a PVAC copolymer emulsion coating) comprises binder and pigment as the main components together with a number of additives which are incorporated to modify properties, much as in compounding thermoplastics and elastomers. Some indeed play the same roles: notably the plasticisers and stabilisers. The pigment may have a decorative function but its presence also improves durability. For paints for exterior use the pigment is a vital line of defence against photo-degradation of the binder. Other additives control properties of the wet paint, such as its flow characteristics, and the stability of the pigment–binder dispersion or, in latex formulations, of the polymer emulsion.

We complete this section by considering briefly the optical and protective functions of the dry pigmented film. A paint coating is frequently required to obliterate the surface appearance of the substrate; the paint technologist speaks of the *hiding power* of a paint as a figure of merit. The hiding power is a measure of the dry film thickness needed to achieve a defined degree of obliteration, usually determined by comparing the reflectance of overpainted black and white test panels. The hiding action of a pigmented film arises from the multiple scattering of incident light passing into it. Incident light is thus returned to the surface, and beyond a certain thickness the paint film is opaque and incapable of transmitting light to and from the surface of the substrate beneath. In the case

TABLE 6.11
Alkyd structure

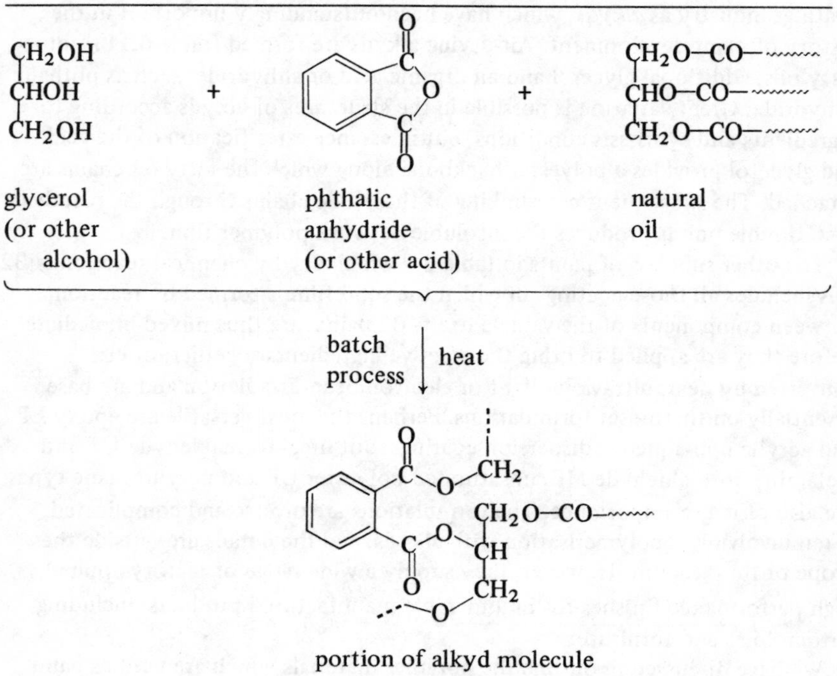

CH₂OH
CHOH   +   +   CH₂O—CO—〰〰
CH₂OH   CHO—CO—〰〰
                                                    CH₂O—CO—〰〰

glycerol   phthalic   natural
(or other   anhydride   oil
alcohol)   (or other acid)

batch   heat
process

portion of alkyd molecule

of a white pigment no absorption occurs and the analysis of optical characteristics
of the film considers scattering events occurring at the interfaces between the
transparent polymer matrix and the embedded transparent pigment particles,
which are typically of the order of 1 μm in diameter. The pigment content of
the dry film may be as high as 60 per cent by volume. The scattering coefficient
σ (*see* section 4.6) can be obtained quite simply by reflectance measurements,
and the paint formulator seeks to improve the hiding power by increasing the
scattering coefficient. The determining factors are the refractive index n of polymer
and pigment materials, and the pigment particle size. Polymer refractive indices
do not vary widely and usually lie close to 1.5 (*see* table 4.3), but more variation
is possible in the selection of the pigment, the refractive index of which should
differ by as much as possible from that of the polymer. A traditional white
pigment such as calcium carbonate (n = 1.63) compares badly with the best
modern white pigment, rutile titanium dioxide (n = 2.76). This difference
underlies the marked difference in the hiding power of paints pigmented with
these two materials — figure 6.12(a). As for particle size, the scattering efficiency
increases with increasing pigment surface area (that is, decreasing pigment

TABLE 6.12

General-purpose emulsion paint formulation
(adapted from B.P. Chemicals (U.K.) Ltd)

	Parts by weight (%)
Vinyl acetate copolymer emulsion*	26.70
Titanium dioxide white pigment	24.50
China clay extender	8.34
Calcium carbonate extender	9.03
Surfactant dispersing aid	0.41
Carboxymethylcellulose colloid thickener, 2 per cent solution in water	10.22
Water	19.73
Fungicide	0.06
Defoaming agent	0.01
Butyl carbitol acetate coalescing solvent (plasticiser)	1.00
	100.00

Total solids content	55 per cent by weight
Pigment volume concentration	50 per cent by volume of dry film

* For example, copolymer of vinyl acetate and vinyl esters of branched carboxylic acids; the emulsion contains 50 per cent by weight of polymer disperse phase in water.

particle size) down to a lower limit which is determined by the wavelength of light (0.4–0.7 μm). Thereafter further reduction in particle size produces a rapid loss of scattering efficiency as refraction and diffraction effects diminish. Figure 6.12(b) illustrates some of these ideas. The need for good control of pigment particle size in hiding paints is obvious, and this means also that the individual particles must be well dispersed in the vehicle, with no tendency for clusters of particles to form.

In addition to hiding and allied decorative functions, many paint films are applied primarily to protect the surface that they cover. What is required of the polymer component in order to achieve this depends very much on the nature of the substrate. Applied to metal surfaces to control corrosion, the important parameters may be permeability to water, oxygen and, above all, dissolved electrolytes. Applied to wood surfaces, the film may have to tolerate relatively high substrate water contents, and withstand severe shrinkage and expansion across the grain as the water content of the wood fluctuates. High permeability to water vapour may be desirable. Applied to concrete surfaces, the film must withstand strongly alkaline conditions and resist hydrolysis. In all circumstances

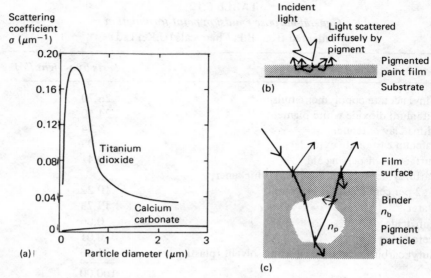

Figure 6.12 (a) Variation of scattering coefficient σ with particle size for paint films with pigments of different refractive index n_p (rutile titanium dioxide 2.76, calcium carbonate 1.63). (b) The scattering of light in paint films: hiding power of a pigmented film arising from diffuse scattering within the film. (c) Ray diagram showing the origin of scattering in reflection and refraction at pigment—binder interfaces. The ratio n_p/n_b determines the extent of ray-splitting and hence scattering efficiency of the pigmented film

the very unfavourable surface to volume ratio of a thin coating (perhaps 50–150 μm thick) makes the film vulnerable to environmental attack, leading to degradation through complex weathering processes including photo-oxidation and hydrolysis (*see* section 5.8).

6.9 Fibres

The defining characteristic of a fibre is its threadlike form, without restriction on composition. In practice nearly all important non-metallic fibres (that is, excluding wires) are polymeric materials, both natural and synthetic. Such natural fibres as wool and cotton have length : diameter ratios of the order of 10^3 or greater. Synthetic fibres are usually produced as continuous filaments. The organic fibres are carbon or heterochain polymers; the major inorganic fibres glass and asbestos are polymeric silicates. In table 6.13 the polymeric fibres of technological importance are broadly classifed according to origin. Table 6.14 classifies the same fibres by primary structure.

TABLE 6.13
Major polymeric fibre materials

Natural fibres	Regenerated natural fibres	Synthetic fibres
Animal	Cellulose (viscose rayon)	*Organic*
Wool	Cellulose acetate CA	PA6, 66, 610 (nylons)
Silk	Cellulose triacetate	Aromatic polyamides
(proteins)		(aramids)
	Natural rubber NR	PE
Vegetable		PP
Cotton		PAN and copolymers
Jute		(acrylics)
(cellulosic)		PVC copolymer
		PVDC copolymer
		Polyurethane
Mineral		elastomer (spandex)
Asbestos		PTFE
(silicate)		PETP (polyester)
		Inorganic
		Carbon
		Graphite
		Glass

TABLE 6.14
Polymeric fibre materials grouped by primary structure

Carbon chain	Heterochain	Silicate
Natural rubber	Wool	Glass
PE	Silk	Asbestos
PP	Cotton	
PAN	Jute	
PVC	Cellulose derivatives	
PVDC	Polyurethane	
PTFE	PA	
Carbon	PETP	
Graphite		

Fibres have two main uses — in the manufacture of textiles and in the production of fibre-reinforced composites (*see* section 6.12); in addition, glass fibres are of growing importance as the main element of optical cables. For all these purposes fibre diameters are almost invariably small, rarely exceeding 25 μm. (For textile fibres, the *linear density* expressed in the units *tex* or *denier* is usually quoted instead of diameter; tex = mg/m and tex = 9 denier.) The value of fibres lies in collective properties which develop when they are combined in large numbers in various fibre assemblies such as yarns, woven, knitted or bonded fabrics, or composites in combination with a variety of matrices. Hence there have evolved several technologies which deal with manufacturing and processing fibres of many kinds for particular end-uses. The details of textile technology and fibre-reinforced plastics technology lie outside the scope of this book but, of course, their products are major outlets for polymer materials.

For most textiles the fibre must possess both a high tensile strength (100–600 MN/m^2) and an extension at break in the range 5–50 per cent — figure 6.13(a). Such mechanical properties in the fibres are necessary for the fabrics produced from them to exhibit the desirable properties of strength and flexibility which are subtly embodied in attributes such as drape. In addition, textile fibres must possess a suitable assortment of chemical characteristics to ensure that they can be dyed, and show reasonable resistance to weathering and other kinds of degradation. The textile technologist pays particular attention to the absorption of water by fibres and to any changes in properties which result from moisture uptake. These changes underlie such characteristics of the fabric as resistance to shrinkage and creasing, and drying speed. The fire performance of textiles is of great practical interest, and finally triboelectric charging is a serious problem both during manufacture and in use.

In practice these diverse requirements are met by a number of crystalline partially oriented polymers, into which class all the leading textile fibres fall. In natural fibres such as cotton and wool the orientation of the polymer chains arises from the organisation imposed during growth by the synthesising enzymes. In cotton each fibre is an individual cell, which contains microfibrils arranged helically about the fibre axis and each comprising about 40 cellulose polymer chains. The cellulose chains are long, with a relative molecular mass of about 5×10^5, corresponding to about 9000 six-atom rings in the chain backbone. They adopt a hydrogen-bonded crystalline arrangement within the microfibrils. The individual fibre of wool in contrast comprises many individual cells and is more highly structured than that of cotton. The principal polymer components are a number of proteins known collectively as keratin. Once again however crystalline microfibrils exist within the fibre, lying parallel to the fibre axis. Within each microfibril the keratin adopts the α-helix conformation found widely in proteins and proposed by Pauling and Corey in 1951.

By contrast, synthetic fibres are produced as continuous filaments, spun either from the melt or from a solution. The initial morphology of the emergent fibre depends on the rate of cooling in melt spinning, on the conditions of

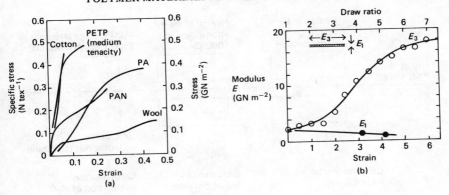

Figure 6.13 (a) Stress—strain curves to failure of typical fibres. The left-hand axis shows specific stress N tex^{-1}, and the right-hand axis shows the corresponding stress assuming that all fibres have density 1000 kg m^{-3}. (b) Mechanical anisotropy in a drawn PETP fibre. E_3 and E_1 are the moduli measured, respectively, along and transverse to the fibre axis. (After Ward)

coagulation or evaporation in spinning from solution, and on the stress in the fibre. The partial orientation of the polymer chains is then achieved by *drawing* the fibre to some 4—5 times its original length. This produces a highly anisotropic fibrillar morphology (*see* chapter 2). The stiffening of the material along the fibre axis is apparent from figure 6.13(b), where the axial and transverse elastic moduli of polyester fibre are shown before and after drawing.

Fibre-reinforced and other composite materials containing polymer components are discussed in the next section. Most commonly the polymer is present as a relatively soft matrix in combination with a relatively high modulus reinforcement, as in glass-fibre reinforced polyester or epoxy composites. Natural fibres such as jute, and also papers, may be used to produce reinforced laminates based on several thermosets, but these reinforcements do not merit any discussion here. The most spectacular high performance polymeric fibre used in producing engineering composites is *carbon fibre* CF. CF is formed from PAN fibres in a controlled pyrolysis process which is outlined in figure 6.14.

The fibre which is intensely black in colour consists of pure carbon with an oriented graphitic structure. CF has an exceptionally high tensile strength and tensile modulus, especially on a weight-for-weight basis, as table 6.15 shows. Indeed the development of CF marks a step forward towards engineering materials approaching the theoretical strengths expected of the primary chemical bonds. Several aramid fibres, which have high modulus, tensile strength and exceptional temperature resistance, rival CF and are finding increasing use in textiles and as reinforcement.

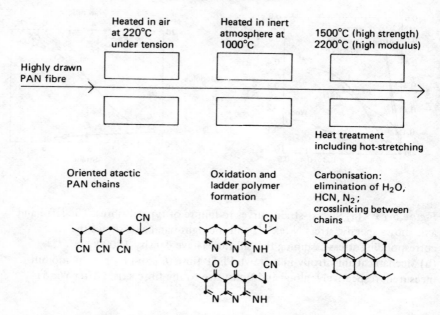

Figure 6.14   Carbon fibre manufacture: outline scheme.

Fibre reinforcement is also a means of improving the properties of brittle materials such as gypsum and concrete. Hair and straw have been added to clay and plaster since the earliest times; more recently asbestos and glass have been added to cements. Polymeric reinforcements, notably fibrillar PP, are also extensively used in concretes and mortars. The function of the fibre is to improve impact resistance.

TABLE 6.15
Mechanical properties of reinforcing fibres

	Young's modulus, E (GN m^{-2})	Tensile strength (GN m^{-2})	Density (kg m^{-3})
Asbestos	190	6	2 500
PA 66	5	0.8	1 100
E glass	60	3.5	2 500
Drawn steel wire	210	4	7 800
Tungsten	345	2.9	19 300
Carbon (high modulus)	420	1.8	2 000
Carbon (high strength)	270	2.8	1 700
Aramid	125	2.9	1 450

6.10 Water-soluble Polymers

Most synthetic polymer materials are insoluble in water and this is usually seen as a virtue. In some circumstances, water solubility can be put to use and some of the main groups of water-soluble polymers are listed in table 6.16. Water-soluble *natural* polymers are numerous. Some degree of chemical bond polarity must be present for a polymer to be soluble in water but the controlling factors are subtle. Thus, for example, polyethylene and polyoxymethylene are insoluble whereas poly(ethylene oxide) is highly soluble. Celluloses are insoluble while starches (which are structurally similar polysaccharides) are soluble. Broadly, carbon chain linear polymers are only water soluble if there are polar oxygen or nitrogen-containing sidegroups attached to about one-half of the C atoms of the main chain. In water, as in other solvents, polymer solubility is frequently limited by phase separation above or below a critical temperature; dissolved polymers are also often precipitated by addition of salts.

Since this book deals with solid polymers, the interest in water-soluble polymers lies in particular applications where the water solubility is exploited either in processing or as an end-use property. For example, polyacrylamide may be chemically crosslinked to form an insoluble gel; the polymer is used in stabilising large soil structures, the polymer being injected as a low viscosity aqueous solution and then converted into a gel *in situ* by a chemically retarded crosslinking reaction. Another example of a polymer material passing through an aqueous solution stage is in the use of poly(acrylic acid)/glass dental cements. The polymer component crosslinks *in situ* by chemical reaction with metal ions leached from the glass filler to form an insoluble network. In contrast, many pharmaceutical products make use of water-soluble polymers, for example in tablet coating and encapsulation, to control the release of drugs in the body.

Most commercial applications of water-soluble polymers are in modifying fluid properties rather than as solid materials.

6.11 Films and Membranes

As films and membranes, synthetic polymer materials have no competitors. Most of the polymer films used today, for example in medicine, packaging and photography, meet needs which could not be met by metals, ceramics or even natural polymers. Several applications provide extremely interesting examples of polymer materials science at work and we discuss two in this section.

Co-extruded Barrier Films

We showed in section 5.2 that the permeability of different polymers to gases and vapours varies greatly. Furthermore the mechanical properties of polymer films also differ greatly. Figure 6.15 illustrates an important innovation in film

TABLE 6.16
Water-soluble polymers and polymer groups

Synthetic carbon chain	Poly(acrylic acid)
	Polyacrylamide
	Poly(methacrylic acid)
	Poly(vinyl alcohol)
	Poly(vinyl methyl ether)
	Polyvinylpyrrolidone
Synthetic heterochain	Poly(ethylene oxide)
	Poly(ethylene imine)
Natural	Polysaccharides
	Guar Gum
	Starch
	Xanthan Gum
	Proteins
	Casein
	Collagen
Chemically-modified natural	Cellulose ethers
	Carboxymethylcellulose

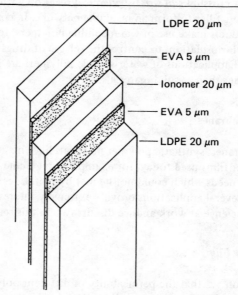

Figure 6.15 Co-extruded multi-layer film: in this example an ionomer barrier layer (with low oxygen permeability) is bonded to LDPE using an EVA bonding agent

technology: the co-extrusion of several different polymer films into a multi-layer composite. Co-extruded composite films are finding wide application in the packaging of food and medical products.

Separation Membranes: Desalination and Dialysis

The two largest uses of polymer membranes for separation of mixtures are in the desalination of water and in medical dialysis. Desalination of seawater and brackish water is commonly carried out by reverse osmosis using cellulose acetate membranes. The CA membrane is permeable to water (via pores a few nm in diameter) but highly impermeable to dissolved salts. In addition to the perm-selectivity, the membrane must be sufficiently strong to resist the pressure applied across it to generate the reverse osmotic flow (at least 20 atm) and be stable against hydrolysis and biological attack. In the case of CA, these two forms of degradation are controlled effectively by operating near neutral pH. In blood dialysis, membranes of regenerated cellulose are used to separate impurities such as urea (which diffuse through the membrane) from macromolecular blood constituents (which do not). In this use, the lack of toxicity and inertness towards biochemically active macromolecules are critical requirements.

6.12 Composites

We have made passing references earlier in this chapter to the importance in polymer technology of combinations of plastics materials with reinforcing fibres and sheet materials. Such combinations are examples of *composite materials*. The mechanics of composite solids has received much attention from engineers and scientists in recent years. The many studies, both theoretical and experimental, of composite behaviour in solids establish a general framework for understanding deformation and fracture in such materials. This emerging theory extends to a much wider range of materials than polymer composites alone, and embraces also materials as diverse as concrete, fibre-reinforced cements and metals, as well as natural substances like wood and bone.

The fibre-reinforced plastics have played an important part in stimulating this activity. The geometrical simplicity of a fibre (or a parallel bundle of fibres) embedded in a polymer matrix was amenable to analysis and this pointed the way to the mechanics of more complex materials. Several texts listed at the end of this chapter provide a comprehensive account of composite mechanics. However, the subject lies beyond the scope of this book.

The important polymer-based composites are numerous. Plywoods, which are laminates of wood veneer bonded with PF and amino resins, are extremely successful commercial materials; PF resins are also used in making particle boards from wood waste. Glass-fibre reinforced unsaturated polyesters and epoxies are the major fibre-reinforced polymer materials, with uses which include both commodity products and high performance engineering.

Suggestions for Reading

Engineering Thermoplastics

Bucknall, C. B., *Toughened Plastics* (Applied Science, London, 1977).
Clagett, D. C., 'Engineering plastics', in *Encyclopaedia of Polymer Science and Engineering*, vol. 6, 2nd edn, pp. 94–131 (Wiley, New York, 1986).
Margolis, J. M. (Ed.), *Engineering Thermoplastics: Properties and Applications* (Dekker, New York, 1985).
Manson, J. A. and Sperling, L. H., *Polymer Blends and Composites* (Heyden, London, 1976).
Ogorkiewicz, R. M. (Ed.), *Thermoplastics: Properties and Design* (Wiley, London, 1974).
Ogorkiewicz, R. M., *The Engineering Properties of Plastics*, Engineering Design Guide no. 17 (Oxford University Press, 1977).
Powell, P. C., *The Selection and Use of Thermoplastics*, Engineering Design Guide no. 19 (Oxford University Press, 1977).
Riess, G., Hurtrez, G. and Bahadur, P., 'Block copolymers', in *Encyclopaedia of Polymer Science and Engineering*, vol. 2, pp. 324–434 (Wiley, New York, 1985).
Rodriguez, F., *Principles of Polymer Systems*, 2nd edn (McGraw-Hill, New York, 1983).
Tess, R. W. and Poehlein, G. W. (Eds), *Applied Polymer Science*, 2nd edn (American Chemical Society, Washington DC, 1985).

Material Property Data

Brandrup, J. and Immergut, E. H. (Eds), *Polymer Handbook*, 2nd edn (Wiley, New York, 1975).
Brydson, J. A., *Plastics*, 4th edn (Butterworth, London, 1982).
Encyclopaedia of Polymer Science and Engineering (Wiley, New York, 1985 onwards) [formerly *Encyclopaedia of Polymer Science and Technology*].
Kroschwitz, J. I. (Ed.), *Polymers: An Encyclopedic Sourcebook of Engineering Properties* (Wiley, New York, 1987).
Modern Plastics Encyclopaedia (McGraw-Hill, New York, *annual*).
Saechtling, H., *International Plastics Handbook* (Hanser, Munich, 1987).

Standards and Testing

ASTM, *Annual Book of ASTM Standards*, Section 8, *Plastics*; Section 9, *Rubbers* (American Society for Testing and Materials, Philadelphia, Pa., *annual*)
BS 2782, *Methods of Testing Plastics* (British Standards Institution, London).
BS 4618, *Recommendations for the Presentation of Plastics Design Data* (British Standards Institution, London).

See also standards issued by the International Organization for Standardization (ISO), Geneva (Technical Committees 61, Plastics; and 45, Rubbers).

Brown, R. P. (Ed.), *Handbook of Plastics Test Methods*, 2nd edn (Godwin, London, 1981).
Brown, R. P., *Physical Testing of Rubbers*, 2nd edn (Elsevier Applied Science, London, 1986).
Brown, R. P. and Read, B. E. (Eds), *Measurement Techniques for Polymeric Solids* (Applied Science, London, 1984).
Kämpf, G., *Characterization of Plastics by Physical Methods* (Hanser, Munich, 1986).
Shah, V., *Handbook of Plastics Testing Technology*, (Wiley, New York, 1984).

Additives

Gaechter, R. and Mueller, H. (Eds), *Plastics Additives Handbook* (Hanser, Munich, 1985).
Mascia, L., *The Role of Additives in Plastics* (Edward Arnold, London, 1974).
Ritchie, P. D. (Ed.), *Plasticizers, Stabilizers and Fillers* (Iliffe, London, 1972).
Sears, J. K. and Darby, J. R., *The Technology of Plasticisers* (Wiley, New York, 1982).

Processing

Becker, W. E., *Reaction Injection Moulding* (Van Nostrand Reinhold, New York, 1979).
Bikales, N. M. (Ed.), *Extrusion and Other Plastics Operations* (Wiley-Interscience, New York, 1971).
Bown, J. and Robinson, J. D., 'Injection and transfer moulding and plastics mould design', *Manual of Light Production Engineering*, vol. 1 (Business Books, London, 1970).
Dubois, J. H. and John, F. W., *Plastics*, 5th edn (Van Nostrand Reinhold, New York, 1974).
Fisher, E. G., *Blow Moulding* (Iliffe, London, 1971).
Fisher, E. G., *Extrusion of Plastics*, 3rd edn (Newnes-Butterworth, London, 1976).
Frados, J. (Ed), *Plastics Engineering Handbook*, 4th edn, Society of the Plastics Industry (Van Nostrand Reinhold, New York, 1976).
Irwin, C., 'Blow molding', in *Encyclopaedia of Polymer Science and Engineering*, vol. 2, 2nd edn, pp. 447–478 (Wiley, New York, 1985).
Kaufman, H. S. and Falcetta, J. F. (Eds), *An Introduction to Polymer Science and Technology*, chs 9–11 (Wiley, New York, 1977).
Kircher, K., *Chemical Reactions in Plastics Processing* (Hanser, Munich, 1987).
Kobayashi, A., *Machining of Thermoplastics* (McGraw-Hill, New York, 1967).

Kruder, G. A., 'Extrusion', in *Encyclopaedia of Polymer Science and Engineering*, vol. 6, 2nd edn, pp. 571–631 (Wiley, New York, 1986).

Meyer, R. W., *Handbook of Pultrusion Technology* (Chapman and Hall, London, 1985).

Middleman, S., *Fundamentals of Polymer Processing* (McGraw-Hill, New York, 1977).

Pearson, J. R. A., *Mechanics of Polymer Processing* (Elsevier Applied Science, London, 1985).

Postans, J. H., *Plastics Mouldings*, Engineering Design Guide no. 24 (Oxford University Press, 1978).

Rubin, I. I., *Injection Moulding Theory and Practice* (Wiley, New York, 1973).

Rubin, I. I., 'Injection molding', in *Encyclopaedia of Polymer Science and Engineering*, vol. 8, 2nd edn, pp. 102–138 (Wiley, New York, 1987).

Tadmor, Z. and Gogos, C. G., *Principles of Polymer Processing* (Wiley, New York, 1979).

Throne, J. L., *Thermoforming* (Hanser, Munich, 1987).

Thermosets

Bruins, P. F. (Ed.), *Unsaturated Polyester Technology* (Gordon and Breach, New York, 1976).

May, C. A. and Tanaka, Y., *Epoxy Resins: Chemistry and Technology* (Dekker, New York, 1973).

Whelan, A. and Brydson, J. A. (Eds), *Developments in Thermosetting Plastics* (Applied Science, London, 1975).

Whitehouse, A. A. K., Pritchett, E. G. K. and Barnett, G., *Phenolic Resins*, 2nd edn (Iliffe, London, 1967).

Elastomers

Barnard, D. *et al.*, 'Natural rubber', in *Encyclopaedia of Polymer Science and Technology*, vol. 12, pp. 178–256 (Wiley, New York, 1970).

Blackley, D. C., *Synthetic Rubbers: their Chemistry and Technology* (Applied Science, London, 1983).

Blow, C. M., and Hepburn, C. (Ed), *Rubber Technology and Manufacture*, 2nd edn (Butterworth, London, 1982).

Cooper, W., 'Synthetic elastomers', in *Encyclopaedia of Polymer Science and Technology*, vol. 5, pp. 406–482 (Wiley, New York, 1966).

Eirich, F. R. (Ed.), *Science and Technology of Rubber* (Academic Press, New York, 1978).

Garvey, B. S. Jr, 'Rubber compounding and processing', in *Encyclopaedia of Polymer Science and Technology*, vol. 12, pp. 280–304 (Wiley, New York, 1970).

Morton, M. (Ed.), *Rubber Technology* (Van Nostrand Reinhold, New York, 1973).

Saltman, W. M., *The Stereo Rubbers* (Wiley, New York, 1977).

Cellular Plastics

Frisch, K. C. and Saunders, J. H. (Eds), *Plastic Foams*, 2 vols (Dekker, New York, 1972).

Shutov, F. A., 'Foamed polymers, cellular structure and properties' in *Advances in Polymer Science*, vol. 51, *Industrial Developments*, pp. 155-218 (Springer, Berlin, 1983).

Shutov, F. A., *Integral/Structural Polymer Foams: Technology, Properties and Applications* (Springer, Berlin, 1986).

Suh, K. W. and Webb, D. D., 'Cellular materials', in *Encyclopaedia of Polymer Science and Engineering*, vol. 3, 2nd edn, pp. 1-59 (Wiley, New York, 1985).

Adhesives

Bikales, N. M. (Ed.), *Adhesion and Bonding* (Wiley–Interscience, New York, 1971).

Kaelble, D. H., *Physical Chemistry of Adhesion* (Wiley–Interscience, New York, 1971).

Kinloch, A. J. (Ed.) *Durability of Structural Adhesives* (Elsevier Applied Science, London, 1983)

Kinloch, A. J. (Ed.), *Structural Adhesives: Developments in Resins and Primers* (Elsevier Applied Science, London, 1986).

Kinloch, A. J., *Adhesion and Adhesives: Science and Technology* (Chapman and Hall, London, 1987).

Shields, J., *Adhesive Bonding*, Engineering Design Guide no. 2 (Oxford University Press, 1974).

Shields, J., *Adhesives Handbook*, 3rd edn (Butterworth, London, 1984).

Skeist, I. (Ed.), *Handbook of Adhesives*, 2nd edn (Van Nostrand Reinhold, New York, 1977).

Wake, W. C., *Adhesion and the Formulation of Adhesives*, 2nd edn (Applied Science, London, 1982).

Coatings

Nylén, P. and Sunderland, E., *Modern Surface Coatings* (Wiley, London, 1965).

Turner, G. P. A., *Introduction to Paint Chemistry*, 2nd edn (Chapman and Hall, London, 1980).

Films and Membranes

Kesting, R. E., *Synthetic Polymeric Membranes: a Structural Perspective*, 2nd edn (Wiley, New York, 1985).

Water-soluble Polymers

Davidson, R. L. (Ed.), *Handbook of Water-Soluble Gums and Resins* (McGraw-Hill, New York, 1980).

Finch, C. A. (Ed.), *Chemistry and Technology of Water-Soluble Polymers* (Plenum, New York, 1983).

Molyneux, P., *Water-Soluble Synthetic Polymers: Properties and Behaviour*, 2 vols (CRC Press, Baton Roca, Fla., 1983).

Fibres and Composites

Berlin, A. A., Volfson, S. A., Enikolopian, N. S. and Negmatov, S. S. (Eds), *Principles of Polymer Composites* (Springer, Berlin, 1986).

Clegg, D. W. and Collyer, A. A. (Eds), *Mechanical Properties of Reinforced Thermoplastics* (Elsevier Applied Science, London, 1986).

Dietz, A. G. H. (Ed.), *Composite Engineering Laminates* (MIT Press, Boston, Mass., 1969).

Friedrich, K., *Fracture Mechanical Behaviour of Short Fiber Reinforced Thermoplastics* (VDI-Verlag, Dusseldorf, 1984).

Gill, R. M., *Carbon Fibres in Composite Materials* (Iliffe, London, 1972).

Hull, D., *Introduction to Composite Materials* (Cambridge University Press, 1981).

Katz, H. S. and Milewski, J. V., *Handbook of Fillers and Reinforcements for Plastics* (Van Nostrand Reinhold, New York, 1978).

Langley, M. (Ed.), *Carbon Fibres in Engineering* (McGraw-Hill, London, 1973).

Lewin, M. and Pearce, E. M. (Eds), *Fiber Chemistry*, Handbook of Fiber Science and Technology, vol. 1 (Dekker, New York, 1985).

Lubin, G. (Ed.), *Handbook of Composites* (Van Nostrand Reinhold, New York, 1982).

McCrum, N. G., *A Review of the Science of Fibre-reinforced Plastics* (Department of Trade and Industry, HMSO, London, 1971).

Morton, W. E. and Hearle, J. W. S., *Physical Properties of Textile Fibres*, 2nd edn (Heinemann, London, 1975).

Nicholls, R. L., *Composite Construction Materials Handbook* (Prentice-Hall, Englewood Cliffs, NJ, 1976).

Parratt, N. J., *Fibre-reinforced Materials Technology* (Van Nostrand Reinhold, London, 1972).

Pritchard, G. (Ed.), *Developments in Reinforced Plastics − 5, Processing and Fabrication* (Elsevier Applied Science, London, 1986).

Sheldon, R. P., *Composite Polymer Materials* (Applied Science, London 1982).

Weatherhead, R. G., *FRP Technology* (Applied Science, London, 1980).

Appendix A: The Names of Polymers and Polymeric Materials

Every newcomer to the study of polymers is inevitably troubled by the profusion of names for individual polymer materials. Some of the difficulties of nomenclature arise directly from the variety and complexity of the materials themselves, but the lack of a universally accepted system of names aggravates the problem.

We must distinguish at least three separate sets of names for polymers and polymer-based materials: (1) the chemical names; (2) the protected commercial or proprietary names; (3) the customary names. In addition, of increasing importance in describing the common plastics and rubbers is a set of standard abbreviations based on widely used chemical names.

Chemical Names

The *systematic* chemical name of a pure polymer is that devised by chemists to describe as fully as possible the constitution of the primary polymer chain. It aims to provide a verbal equivalent of the symbolic chemical formula. Any chemical substance can be assigned such a name provided that its molecular structure is known; the rules for constructing the name form a complex nomenclature system.

Unfortunately in the polymer field many of the widely used chemical names of important polymers are at variance with systematic nomenclature (and some indeed are thoroughly misleading). At present the deliberate use of systematic naming is largely confined to the research literature.

In fact the chemical names of most common polymers have been based on the name of the source material according to the following simple rules.

For homopolymers	Monomer name	Polymer name
	X	polyX
	XY	poly(XY)
for example	ethylene	polyethylene
or	methylpentene	polymethylpentene
	vinyl chloride	poly(vinyl chloride)
	ω-aminocaproic acid	poly(ω-aminocaproic acid) [nylon 6]

Where the monomer name consists of two words, it should be bracketed in the polymer name to avoid ambiguity. Unfortunately, this practice is frequently not followed.

For copolymers	Monomer name		Polymer name
	X, Y		X–Ycopolymer
		or	poly(X-co-Y)
for example	acrylonitrile, styrene		acrylonitrile–styrene copolymer
		or	poly(acrylonitrile-co-styrene)

It is clear that a name based on the starting material is on the whole better suited to addition than to condensation polymers. Several condensation polymers have been named directly from the structure of the repeat unit of the polymer chain, thus

Repeat unit name	Polymer name
X	polyX
oxymethylene	polyoxymethylene

Common polyesters and polyamides follow this form, although the naming of the repeat unit leaves something to be desired

    polyesters, e.g. poly(ethylene terephthalate)
    polyamides, e.g. poly(hexamethylene adipamide) [nylon 66]

The important classes of heterochain polymers, such as the epoxies, polyesters, polyamides and polyurethanes, each contain many chemically complex individuals and have their own intricate naming usages. The most widely known designation is usually the chemical class name, based on the characteristic chemical unit common to all members of the class. For example

epoxy	epoxy group
polyamide	amide group
polyester	ester group
polyurethane	urethane group

In the case of the polyamides, the term nylon is an accepted class name. Specific members of the class are identified by a number which indicates the number of carbon atoms in the repeat unit. A single number refers to an aminoacid polymer (for example, nylon 6); a two-part number (for example, nylon 66) indicates the numbers of carbon atoms contributed, respectively, by the diamine and the dibasic acid.

Widely used chemical names of important materials are listed in table A1, column 2.

TABLE A.1

Abbreviation	Polymer-based material	Customary names
ABS	acrylonitrile—butadiene—styrene	
CA	cellulose acetate	acetate
CAB	cellulose acetate butyrate	butyrate
CF	cresol–formaldehyde	
CMC	carboxymethylcellulose	
CN	cellulose nitrate	celluloid, nitrate
CPE	chlorinated polyethylene	
CPVC	chlorinated poly(vinyl chloride)	
CS	casein	
EC	ethyl cellulose	
EP	epoxy, epoxide	
EVA	ethylene-vinyl acetate copolymer	
FEP	fluorinated ethylene-propylene copolymer	
MF	melamine–formaldehyde	melamine
PA	polyamide	nylon
PAN	polyacrylonitrile	
PBTP	poly(butylene terephthalate)	
PC	polycarbonate	
PCTFE	polychlorotrifluoroethylene	
PE	polyethylene	polythene
PEBA	polyether block amide	
PEEK	polyetheretherketone	
PEI	polyetherimide	
PEO	poly(ethylene oxide)	

TABLE A1 continued

Abbreviation	Polymer-based material	Customary names
PETP or PET	poly(ethylene terephthalate)	polyester
PF	phenol–formaldehyde	phenolic
PI	polyimide	
PIB	polyisobutylene	
PMMA	poly(methyl methacrylate)	acrylic
PMP	polymethylpentene	
POM	polyoxymethylene	acetal, polyacetal, polyformaldehyde
PP	polypropylene	
PPS	poly(phenylene sulphide)	
PPSU	poly(phenylene sulphone)	
PS	polystyrene	styrene
PSU	polysulphone	
PTFE	polytetrafluorethylene	
PUR	polyurethane	
PVAC	poly(vinyl acetate)	
PVAL	poly(vinyl alcohol)	
PVB	poly(vinyl butyral)	
PVC	poly(vinyl chloride)	vinyl
PVDC	poly(vinylidene chloride)	
PVDF	poly(vinylidene fluoride)	
PVF	poly(vinyl fluoride)	
SAN	styrene–acrylonitrile copolymer	
SB	styrene–butadiene copolymer	
SI	silicone	
UF	urea–formaldehyde	urea
UP	unsaturated polyester	polyester

Elastomers

ACM	polyacrylate	acrylic
ANM	acrylate–acrylonitrile copolymer	
AU	polyester–urethane	polyurethane
BR	polybutadiene	
CM	chlorinated polyethylene	
CO	polychloromethyloxirane (epichlorhydrin polymer)	polyether
CR	polychloroprene	neoprene
CSM	chlorosulphonated polyethylene	
ECO	epichlorhydrin copolymer	polyether
EPDM	ethylene–propylene–diene terpolymer	

TABLE A.1 continued

Abbreviation	Polymer-based material	Customary names
EPM	ethylene—propylene copolymer	
EU	polyether—urethane	polyurethane
FKM	carbon chain fluoropolymer	fluorocarbon
FVMQ	fluorosilicone rubber (vinyl, methyl)	
GPO	propylene oxide copolymer	
IIR	isobutene—isoprene copolymer	butyl
IR	synthetic polyisoprene	
MQ	silicone elastomer (methyl)	
NBR	acrylonitrile—butadiene copolymer	nitrile
NR	natural rubber	
PMQ	silicone elastomer (phenyl, methyl)	
PMVQ	silicone elastomer (phenyl, vinyl, methyl)	
SBR	styrene—butadiene copolymer	
VMQ	silicone elastomer (vinyl, methyl)	
YSBR	thermoplastic styrene—butadiene copolymer	

ASTM Elastomer Classes

M	rubbers having a saturated carbon chain
O	heterochain polymers having O in the chain
R	unsaturated carbon chain rubbers
Q	heterochain polymers having Si in the chain
T	heterochain polymers having S in the chain
U	heterochain polymers having C, O and N in the chain

Commercial or Proprietary Names

These names are created, usually as trade marks, by the primary materials manufacturers; it must be emphasised that they refer to commercial products or ranges of products. Such materials are rarely pure polymers, and the commercial name may therefore imply a certain formulation of base polymer and additives. It may also carry a specification of mechanical and other properties. The complete commercial name of a plastic or rubber material may therefore provide through the manufacturer a very full identification. However, such a name is not intrinsically informative.

Many commercial names are devised according to house rules which may be helpful in identifying the manufacturer.

Customary Names

A miscellany of names of long standing has become widely established. Some of these (such as celluloid, nylon and polythene) were originally proprietary names which over the years have lost their claim to protection. Others such as acetal and acrylic are inaccurate and uninformative chemical names. There is little to be said in favour of most of these customary names and with few exceptions their use is to be discouraged. Some common customary names are listed in column 3 of table A1.

Abbreviations

The general use of systematic or at least of unambiguous chemical names would bring great advantages. In recent years this has been assisted by the adoption of standard abbreviations of the chemical names of the most important polymeric materials. These abbreviations are rapidly gaining favour. Those given in table A1 are taken from the lists of ASTM, BSI, ISO and IUPAC. They should be used as printed, in capitals, without punctuation.

Sources

ASTM D1600–86, *Standard abbreviations of terms relating to plastics* (American Society for Testing and Materials, Philadelphia, Pa.).

ASTM D1418–85, *Rubber and rubber latices – nomenclature* (American Society for Testing and Materials, Philadelphia, Pa.).

BS 3502 : 1978, *Schedule of common names and abbreviations for plastics and rubbers.*

Fox, R. B., 'Nomenclature', in *Encyclopaedia of Polymer Science and Technology*, vol. 9, pp. 336–344 (Wiley, New York, 1968).

International Union of Pure and Applied Chemistry, 'List of standard abbreviations (symbols) for synthetic polymers and polymer materials', *Pure Appl. Chem.*, **40** (1974) 475–476.

ISO 1043–1978, *Plastics – Symbols* (International Organization for Standardization, Geneva).

ISO 1629–1976, *Rubber and Latices – Nomenclature* (International Organization for Standardization, Geneva).

See also:

ASTM D883–86, *Standard definitions of terms relating to plastics* (American Society for Testing and Materials, Philadelphia, Pa.).

ASTM D1566–87a, *Standard terms relating to rubbers* (American Society for Testing and Materials, Philadelphia, Pa.).

Rubber and Plastics Research Association (RAPRA), *New trade names in the rubber and plastics industries* (annual).

Appendix B: Glossary of Polymer Materials

This glossary emphasises the *individuality* of polymers and polymer families, historically, scientifically and technically. It complements the rest of the book, in which polymer materials as a class are treated *as a whole* and by *property*. Students may find the glossary useful in testing and reinforcing their knowledge of much of the material covered in chapters 1–6. Selected references to the literature of individual polymers are included.

Acetal
See Polyoxymethylene

Acrylate elastomers
A small group of carbon-chain synthetic elastomers (ASTM class M) based on the poly(ethyl acrylate) linear chain copolymerised with a small amount of comonomer to allow chemical crosslinking. Acrylate rubbers have outstanding resistance to oxidative degradation.
Hagman, J. F. and Crary, J. W., 'Acrylic elastomers', in *Encyclopaedia of Polymer Science and Engineering*, vol. 1, 2nd edn, pp. 306–334 (Wiley, New York, 1985).

Acrylic ester polymers
A group of amorphous carbon-chain polymers of which by far the most important as a thermoplastic is poly(methyl methacrylate) PMMA. PMMA (customary name: acrylic) is a colourless, highly transparent glassy polymer with outstanding resistance to weathering. Its exceptional durability and resistance to attack by inorganic substances is somewhat offset by its solubility in a variety of organic solvents. However, dissolving PMMA or other acrylate ester polymers in a well matched solvent such as methyl ethyl ketone (PMMA 9.4; MEK 9.3) produces a coating lacquer. Acrylics are in fact among the most important industrial

coating resins, especially for motor vehicles. Many acrylic coatings are thermo-sets, in which a reactive comonomer is introduced into the polyacrylate chain to allow crosslinking, for example, by epoxy or urethane chemistry. In recent decades, pollution and safety pressures have driven the acrylic coatings industry to develop water-based emulsion (latex) systems, and high-solids or powder systems. The acrylic ester polymers illustrate nicely the variation of glass trans-ition temperature with primary chain structure:

poly(methyl methacrylate)	105°C
poly(ethyl methacrylate)	65
poly(n-propyl methacrylate)	35
poly(n-butyl methacrylate)	20
poly(methyl acrylate)	9
poly(ethyl acrylate)	−22
poly(n-propyl acrylate)	−48
poly(n-butyl acrylate)	−54

Qualitatively, this trend shows that the bulkier side groups keep adjacent chains somewhat further apart, thus making segmental chain motion easier at any particular temperature. This systematic dependence on chain structure is exploited to produce coating formulations of different hardness or flexibility for different uses and service temperatures. Lightly crosslinked poly(ethyl acrylate) has a sufficiently low glass transition temperature to be a successful elastomer (*see Acrylate elastomers*).

Acrylonitrile–butadiene–styrene

The first of the ABS resins was brought into production by Borg–Warner in 1954. ABS is a hybrid thermoplastic in which polybutadiene rubber particles are dis-persed in a poly(styrene-*co*-acrylonitrile) SAN matrix. The dispersed rubber phase gives ABS a much greater impact strength than SAN itself. The main technical achievement in developing ABS lay in stabilising the BR dispersion by chemically grafting the SAN copolymer at the particle surface. Recently, ABS has itself been blended with PVC, with polycarbonate, with polyamides and with polysulphone to produce further hybrid polymer materials.

The ABS/PVC hybrid is a four-component engineering polymer which has evolved by a long development path from the homopolymers polystyrene, polyacrylonitrile, polybutadiene and poly(vinyl chloride). There are great differences between the parent homopolymers: PS glassy and brittle, PAN crystalline and fibre-forming, polybutadiene rubbery. Simple copolymerisation allows some modification of properties. Each of the three possible two-com-ponent copolymers is an important commercial material: SAN as an improved polystyrene; and the two synthetic rubbers, SBR and NBR. Each of these is a single-phase random copolymer material. In the two-phase three-component material ABS, the hybrid acquires some of the attributes of each of the phases and control of the morphology brings some degree of microstructural engineering to the development of new materials. In the four-component hybrids yet another composition variable is introduced.

Annual production capacity for ABS worldwide is about 1 million tonnes.

Kulich, D. M., Kelly, P. D. and Pace, J. E., 'Acrylonitrile–butadiene–styrene polymers', in *Encyclopaedia of Polymer Science and Engineering*, vol. 1, 2nd edn, pp. 388–426 (Wiley, New York, 1985).

Alkyd resins

Alkyds are long-used traditional resins of the paint industry, essentially polyesters, frequently modified with natural drying oils. Today, they are often blended with other polymer types such as cellulosics and phenolics. The alkyds remain the most important of the surface coating resins (*see* page 185).

Lanson, H. J., 'Chemistry and technology of alkyd and saturated reactive polyester resins', in R. W. Tess and G. W. Poehlein (Eds), *Applied Polymer Science*, 2nd edn, pp. 1181–1204 (American Chemical Society, Washington DC, 1985).

Amino resins

The amino thermosets are based on the reaction of melamine or urea with formaldehyde. The manufacture and processing is similar to that of the phenolics. They have an unspectacular but safe niche in the polymer materials market, finding innumerable uses as adhesives, moulded products and laminates. Among these, however, the largest volume by far goes to the wood products industry as a binder for particle board. UF resins can be combined with blowing agents to produce foams, especially for *in situ* thermal insulation. There has been some concern since the mid 1970s about the toxicity of residual formaldehyde vapour.

Updegraff, I. H., 'Amino resins', in *Encyclopaedia of Polymer Science and Engineering*, vol. 1, pp. 752–789 (Wiley, New York, 1985).

Vale, C. P. and Taylor, W. G. K., *Amino Plastics* (Iliffe, London, 1964).

Williams, L. L., Updegraff, I. H. and Petropoulos, J. C., 'Amino resins', in R. W. Tess and G. W. Poehlein (Eds), *Applied Polymer Science*, 2nd edn, pp. 1101–1115 (American Chemical Society, Washington DC, 1985).

Aramid

Aromatic polyamides (*see* table 5.7 for structure) with high temperature stability and exceptional mechanical properties as fibres; discovered by Stephanie Kwolek at Du Pont in 1968. The aramid chains are relatively rigid and form liquid crystalline solutions from which the polymer may be produced as a highly oriented fibre by spinning.

Preston, J., 'Aramid fibers', in M. Grayson (Ed.), *Encyclopaedia of Composite Materials and Components*, pp. 97–126 (Wiley, New York, 1983).

Butyl rubber

Butyl rubber IIR, a synthetic elastomer made by copolymerising isobutylene with small amounts of isoprene, was brought into commercial production by Standard Oil in 1943. The isoprene contributes the crosslinking sites to an otherwise saturated polyisobutylene chain. After vulcanising, the residual unsaturation of IIR is very small and this leads to good resistance to oxidative degradation. IIR (with its halogenated forms, chlorobutyl and bromobutyl rubber) has about 5 per cent of the total rubber market. The largest application of IIR is in making tyre inner tubes, where its low gas permeability and oxidative stability are valuable.

Kresge, E. N., Schatz, R. H. and Wang, H.-C., 'Isobutylene polymers', in *Encyclopaedia of Polymer Science and Engineering*, vol. 8, 2nd edn, pp. 423–448 (Wiley, New York, 1987).

Carbon fibre

A polymeric fibre produced from polyacrylonitrile by controlled thermal oxidation and used in high performance composites (see page 191).

Riggs, J. P., 'Carbon fibres', in *Encyclopaedia of Polymer Science and Engineering*, vol. 2, 2nd edn, pp. 640–685 (Wiley, New York, 1985).

Cellulose and its derivatives

Cellulose itself is a natural polymer which plays an essentially structural role in plant tissue. It is found (in slightly different molecular forms) in cotton and in wood. Natural cellulose is a highly crystalline linear polymer (built from glucose units) which is generally found in a fibrillar morphology in biological materials. The relative molecular mass is typically high, 0.5–1.5×10^6. Natural cellulose fibres such as cotton are of course used directly, but apart from that the chemical technology of cellulose depends on chemical modification since cellulose is insoluble. The earliest modified cellulose was cellulose nitrate (celluloid), which despite its flammability is still used, primarily as film and in coatings. Cellulose acetate (and other organic esters such as the closely related cellulose acetate-butyrate CAB) are less flammable and of greater importance, mainly in the coatings industry as high-quality lacquers. These cellulosics are insoluble in water. In contrast, another group of cellulose derivatives, the cellulose ethers, are water-soluble and are among the most important of the water-soluble polymers as adhesives and thickeners.

Epichlorhydrin rubber

A minor special-purpose polyether elastomer (ASTM class O), somewhat resembling nitrile rubber with excellent oil resistance but rather better heat and ozone resistance; commercialised in 1970.

EPM/EPDM rubbers (ethylene–propylene rubbers)

An increasingly significant group of synthetic rubbers based on ethylene-propylene random copolymers. EPM is vulcanised by peroxides; EPDM with sulphur through a small amount of diene incorporated into the linear chains. These ethylene–propylene rubbers are very resistant to oxidative degradation. They hold about 5 per cent of the synthetic rubber market. EPDM is alloyed with polypropylene to form a thermoplastic elastomer (*see Thermoplastic elastomers*).

Corbelli, L. 'Ethylene–propylene rubbers', in A. Whelan and K. S. Lee (Eds), *Developments in Rubber Technology*, vol. 2, pp. 87–129 (Applied Science, London, 1981).

Strate, G. V., 'Ethylene–propylene elastomers', in *Encyclopaedia of Polymer Science and Engineering*, vol. 6, 2nd edn, pp. 522–564 (Wiley, New York, 1986).

Epoxy resins

A large, chemically complicated class of thermosetting resins of great value as adhesives, coatings and in composites. *See* page 30 for an outline of epoxy chemistry. The epoxies have remarkable resistance to chemical attack and combine adhesion and good mechanical properties (impact resistance, toughness, elastic modulus and strength). Their existence has therefore greatly stimulated the development of high-performance structural adhesives and composites, for example in aerospace and construction engineering.

Lee, H. and Neville, K., *Handbook of Epoxy Resins* (McGraw-Hill, New York, 1967).

McAdams, L. V. and Gannon, J. A., 'Epoxy resins' in *Encyclopaedia of Polymer Science and Engineering*, vol. 5, 2nd edn, pp. 322–382 (Wiley, New York, 1985).

Ethylene–propylene rubbers
See EPM/EPDM rubbers

Fluorocarbon polymers

Replacement of hydrogen by fluorine in the C–H bond produces an increase in chemical stability: compare PE and PTFE. PTFE is the leading member of the fluorocarbon group which now also includes a considerable number of fluorinated thermoplastics and elastomers with small, specialist market niches. Besides PTFE, the fully fluorinated thermoplastics include PCTFE and the fluorinated ethylene–propylene copolymer FEP. The partially fluorinated poly(vinyl fluoride) PVF and poly(vinylidene fluoride) PVDF are less stable than their fully fluorinated counterparts but more stable than their chlorinated analogues. The fluorocarbon elastomers fall into the ASTM class M: FFKM designating fully fluorinated materials and FKM the partially fluorinated types.

Interpenetrating Polymer Networks (IPNs)

A further variation on the theme of polymer hybridisation. The IPN concept (mainly of scientific interest so far) envisages a polymer X (or its monomer) penetrating into an already crosslinked polymer network Y and then itself becoming crosslinked. Ideally, the resultant XY hybrid is single-phase, the two networks being inextricably interwoven. In practice, some phase separation usually occurs but the domain size is small and may be controlled by the crosslink density. Numerous IPNs have been produced for scientific study but few are produced commercially. One example is the blending of phenolic thermosetting resins with SBR and NBR synthetic rubbers. The PF novolak with crosslinker is blended with the elastomer before vulcanisation. No chemical reaction between phenolic resin and elastomer occurs; the interpenetrating phenolic network acts to stiffen and harden the rubber.

Klempner, D. and Berkowski, L., 'Interpenetrating polymer networks', in *Encyclopaedia of Polymer Science and Engineering*, vol. 8, 2nd edn, pp. 279–341 (Wiley, New York, 1987).

Ionomer polymers

A group of linear chain thermoplastics containing up to 20 mol per cent (but usually considerably less) of an acid monomer which is neutralised by a metal or quaternary ammonium ion, thus achieving strong interchain bonding. Du Pont introduced the first major material of this type in 1964, poly(ethylene-*co*-methacrylic acid). It may be regarded as a modified polyethylene, having excellent optical clarity and better tensile properties than LDPE. There is evidence that the ionic groups in ionomers tend to aggregate into clusters giving an inhomogeneous microstructure. Du Pont later introduced another group of ionic polymers, the Nafion resins. These may be regarded as modified PTFEs, in which tetrafluorethylene is copolymerised with sulphonated fluoralkoxy comonomers. The polymer structure is therefore a heavily fluorinated linear polymer with pendant fluoralkoxy groups terminating in a sulphonate salt group. The sulphonate group reversibly binds a metal ion and Nafion resins are used as *ion-exchange resins* in water purification. In addition, at least one thermoplastic elastomer is ionic, the sulphonated EPDM.

Longworth, R., 'Structure and properties of ionomers', in A. D. Wilson and H. J. Prosser (Eds), *Developments in Ionic Polymers*, vol. 1, (Applied Science, London, 1983).

Lundberg, R. D., 'Ionic polymers', in *Encyclopaedia of Polymer Science and Engineering*, vol. 8, 2nd edn, pp. 393–423 (Wiley, New York, 1987).

Liquid crystalline polymers

Linear polymers with a stiff primary chain structure may spontaneously align their molecular axes in solution or in the melt to form liquid crystals. Alignment is enhanced in shear flow and much of the interest in these materials arises from their unusual flow properties. Thus they have very low melt viscosities because of the lack of chain entanglement, and may be spun from solution or from the melt to give fibres with exceptional strength and modulus. A high degree of chain alignment occurs in the extensional flow which occurs during spinning.

Aramids are liquid crystalline in solution (lyotropic) — see *Aramid*; a number of aromatic polyesters are liquid crystalline in the melt (thermotropic). None of the thermotropic polyesters is yet well established commercially.

Dobb, M. G. and McIntyre, J. E., 'Properties and applications of liquid-crystalline main-chain polymers', *Adv. Polymer Sci.*, **60/61** (1984) 61–98.

Natural rubber

The industrial use of natural rubber pre-dates the synthetic polymer industry by a hundred years. The major synthetic rubbers were developed several decades ago and the balance between NR and the synthetics is surprisingly stable: NR 40 per cent, synthetics 60 per cent. Vehicle tyres are the dominant market for both, taking about 60 per cent of total rubber output. NR remains the supreme general-purpose elastomeric material, its sensitivity to solvents and chemical/atmospheric degradation being its main shortcomings. Scientifically, the study of natural rubber has made several important contributions to polymer science. First, the rapid deterioration of unstabilised rubber exposed to the atmosphere stimulated a detailed molecular study of the role of oxygen and ozone in chain degradation. Subsequently, stabilisers were developed to control the reactivity of the polymer. Second, the high elasticity of natural rubber attracted much scientific attention (including that of Staudinger, Mark and Flory), from which the theory of rubber elasticity eventually emerged. This theory is one of the cornerstones of polymer science itself and contains within it much of the basic physics of the polymer chain and the polymeric network. Third, the very precise stereoregularity of natural rubber (essentially stereo-chemically perfect poly(*cis*-1,4-isoprene)) set down a challenge for chemists working on synthetic rubbers. In due course, the Ziegler–Natta catalysts pro-vided a tool with which it became possible to achieve extraordinary steric control and a number of synthetic stereo-rubbers were ultimately developed. (*See* page 58, rubber elasticity; page 167, elastomer technology.)

Barlow, C., *The Natural Rubber Industry: its Development, Technology and Economy in Malaysia* (Oxford University Press, 1978).

Elliott, D. J., 'Developments with natural rubber' in A. Whelan and K. S. Lee (Eds), *Developments in Rubber Technology*, vol. 1, pp. 1–44 (Applied Science, London, 1979).

Roberts, A. D. (Ed.), *Natural Rubber Science and Technology* (Oxford University Press, 1988).

Nitrile rubbers

NBR, poly(butadiene-*co*-acrylonitrile) elastomer, is one of the two important synthetic rubbers developed in Germany around 1934 (the other being SBR). NBR now holds less than 5 per cent of the world synthetic rubber market but has many important uses because of its outstanding resistance to petroleum solvents. The acrylonitrile content may vary from about 10 to 40 per cent, more acrylonitrile bringing improved solvent resistance but with some loss of low-temperature flexibility. NBR forms a single-phase blend with PVC.

Bertram, H. H., 'Developments in acrylonitrile–butadiene rubber (NBR) and future prospects', in A. Whelan and K. S. Lee (Eds), *Developments in Rubber Technology*, vol. 2, pp. 51–85 (Applied Science, London, 1981).

Nylons
See Polyamides

Phenolic resins
The family of thermosetting resins of which phenol–formaldehyde was the first (and remains the most important) example, discovered by Baekeland in 1907. For the manufacture and processing of these resins, *see* page 165.

Knop, A. and Pilato, L. A. (Eds), *Phenolic Resins, Chemistry, Applications, and Performance* (Springer, Berlin, 1985).

Polyacetylene
A hydrocarbon polymer of considerable research interest which is produced with the aid of a Ziegler catalyst from acetylene gas, C_2H_2

$$n\ CH\equiv CH \quad \rightarrow \quad +CH=CH+_n$$

The polymer has alternating double bonds in the main chain and has been intensively studied in the search for polymers with high intrinsic electrical conductivity. Stereoregular *trans*-polyacetylene is a semi-conductor with a conductivity of about 10^{-3} $(ohm\ m)^{-1}$. Its conductivity, however, is enormously increased by doping with small quantities of electron-donor or acceptor species, such as iodine. Iodine-doped *trans*-polyacetylene has metallic conductivity of 10^5 $(ohm\ m)^{-1}$. It was also discovered that polyacetylene could be polymerised in the solid state by inducing the molecules in crystalline diacetylene to combine, thus producing very perfect fibrous polymer single crystals.

Chien, J. C. W., *Polyacetylene: Chemistry, Physics and Materials Science* (Academic Press, New York, 1984).

Young, R. J., 'Polymer single crystal fibres', in I. M. Ward (Ed.), *Developments in Oriented Polymers*, vol. 2 (Elsevier Applied Science, London, 1987).

Polyacrylamide
A water-soluble linear carbon-chain polymer which is a close structural relation of the acrylic ester and acrylic acid polymers. It is not used as a solid material but very widely employed in solution and in gel form, as a viscosifier and thickening agent. Acrylamide may be copolymerised with a suitable bi-functional comonomer to produce a crosslinked gel. The remarkable swelling properties of such gels have been the subject of a number of studies by Tanaka (*see* page 125).

Polyamides (or Nylons)
The nylons are indelibly associated with the name of Wallace H. Carothers of Du Pont who carried out the classic studies of step reaction polymerisation in

the late 1920s and early 1930s which led to the commercial development of nylon-66. (Carothers' work was enormously influential in the development of polymer science because he used well known reactions of conventional organic chemistry to construct novel polymeric structures. At a time when the polymer concept was controversial, his work convinced many sceptics of the *reality* of the polymer molecule.) Nylon-6 was developed by Paul Schlack at IG Farben-industrie in Germany between 1937 and 1940. Worldwide, nylon-66 and nylon-6 are produced in roughly equal quantities and together account for about 90 per cent of total nylon output. Other commercial materials are nylon-610, nylon-9, nylon-11 and nylon-12.

Polyamides are tough crystalline polymers which have wide application as fibres and engineering thermoplastics. First applications were as fibres but in the 1950s Du Pont pioneered the idea of 'engineering thermoplastics' with their nylon moulding compounds. Subsequently, nylons reinforced with short glass fibres were successfully introduced, providing greater stiffness than unreinforced forms. More recently, polyamides have become important as components in hybrid materials: for example, in block copolymers with polyethers such as poly(ethylene glycol) to produce *polyether block amides*; or in alloys with poly(phenylene oxide). In the 1970s, a new group of aromatic polyamides, the aramids, were developed with remarkable fibre strength and temperature resist-ance (*see* pages 136–137 and 188–192).

Polyarylates

A wholly aromatic engineering thermoplastic introduced by Unitika (Japan) in 1974. The linear heterochain structure of this type

yields a tough, glassy material (T_g around 200°C); with good high temperature performance.

Polybenzimidazole

For molecular structure, *see* table 5.7. Polybenzimidazole is produced com-mercially in small quantities by Celanese mainly for applications which make use of its outstanding high-temperature performance. Both the heat deflection temperature and the glass transition temperature are quoted as 435°C. The limiting oxygen index is around 40. Exposed to high temperatures in air, the polymer does not melt but converts slowly to a carbon char without flaming. It retains some useful strength at temperatures as high as 650°C. Remarkably, polybenzimidazole can be spun into fibres from a concentrated solution of prepolymer. The fibre proves to be quite suitable for weaving (elongation to break of 30 per cent, tenacity about 3 g/denier), opening the way to applications in heat-resistant polymeric textiles.

Powers, E. J. and Serad, G. A. 'History and development of polybenzimidazoles', in R. B. Seymour and G. S. Kirshenbaum (Eds), *High Performance Polymers: their Origin and Development* (Elsevier Applied Science, London, 1986).

Polybutadiene

Butadiene monomer is one of the polymer industry's key monomers, being used in ABS thermoplastics, in the SBR and NBR synthetic rubbers, in styrene-butadiene copolymer thermoplastic elastomers and in latices. Cheap butadiene and Ziegler–Natta catalysts stimulated the development of stereoregular homopolymer polybutadienes (both *cis* and *trans*). Polybutadiene BR now ranks second among the synthetic rubbers (world capacity about 1.75 million tons per year, roughly 15 per cent of total synthetic rubber). BR is almost entirely used in physical blends with SBR and NR, mainly for tyres; and also as the rubber phase in HIPS (*see* figure 6.1).

Tate, D. P. and Bethea, T. W., 'Butadiene polymers', *Encyclopaedia of Polymer Science and Engineering*, vol. 2, 2nd edn, pp. 537–590 (Wiley, New York, 1985).

Polybutene

The homopolymer is a very minor member of the polyolefin family, produced commercially only in small quantities. However 1-butene is an important co-monomer in the production of high-density polyethylene HDPE and linear low-density polyethylene LLDPE.

Chatterjee, A. M., 'Butene polymers', in *Encyclopaedia of Polymer Science and Engineering*, vol. 2, 2nd edn, pp. 590–605 (Wiley, New York, 1985).

Rubin, I. D., *Poly(1-butene)* (Gordon and Breach, New York, 1968).

Poly(butylene terephthalate)

A thermoplastic polyester currently enjoying very rapid market growth. World consumption of PBTP is over 100 000 tonnes (1987), increasing each year by about 20 per cent. Its material properties are broadly similar to those of its older sibling PETP. However, in injection moulding, it has an important advantage over PETP, which is very slow to crystallise (so that cycle times are long and indeed long-term crystallisation shrinkage may occur). PBTP crystallises extremely easily and rapidly and has now established itself as an almost ideal injection moulding thermoplastic. It is also one of the most versatile thermoplastics for use in polymer hybrids, notably with PC and PETP.

Polycarbonate

A linear heterochain polymer (*see* table 1.5 for structure), developed simultaneously by Bayer and General Electric. Structurally, it is a condensation product of bisphenol A and carbonic acid, hence a carbonate. PC is usually regarded as the second of the five major engineering thermoplastics (PA, PC,

POM, PPO/PS and the thermoplastic polyesters PETP, PBTP) with an estimated world market of about 400 000 tons in 1986. Polycarbonates are tough, transparent materials. PC is increasingly valuable as a component in polymer hybrids, notably PC/ABS and PC/PBTP.

Polychloroprene

Poly(2-chlorobutadiene), a long-established synthetic rubber, discovered by W. H. Carothers who is more widely known for his work on polyamides, now having less than 5 per cent of the total synthetic rubber market. CR has good oil resistance but it is now challenged by many new special purpose elastomers.

Stewart, C. A., Jr, Takeshita, T. and Coleman, M. L., 'Chloroprene polymers', in *Encyclopaedia of Polymer Science and Engineering*, vol. 3, pp. 441–462 (Wiley, New York, 1985).

Polyesters

The broad family name for the heterochain polymer materials with the ester $-CO-O-$ sequence in the main chain. The polyester family includes the saturated polyesters, poly(ethylene terephthalate) PETP and poly(butylene terephthalate) PBTP; the unsaturated polyester thermosets used in glass-fibre resin and composite technology (*see* page 29); the alkyds and drying oils used in surface coatings (*see* page 184); and the new polyarylate engineering thermoplastics (*see Polyarylates*). In addition, polyester blocks are incorporated in polyurethanes and thermoplastic elastomers (*see* page 173).

Meyer, R. W., *Handbook of Polyester Molding Compounds and Molding Technology* (Chapman and Hall, New York, 1987).

Polyetherester elastomers

A term used to describe the block copolymer thermoplastic elastomers with ASTM designation YBPO (polyether soft block, aromatic polyester hard block). These materials have oil resistance similar to nitrile rubber and better heat resistance.

Polyetheretherketone

PEEK was developed by ICI and first offered commercially in 1978. It is one (*see* table 5.7 for structure) of a group of linear heterochain polymers with benzene ring (aromatic) units in the chain (others being poly(phenylene sulphide) and the polysulphones). The melting temperature of PEEK is around 330°C (T_g 145°C). It has outstanding thermal stability with a UL thermal index of about 240°C, the highest of all the melt processable thermoplastics. The chain structures of PES and PEEK are very similar, but PES is amorphous while PEEK is crystalline.

Polyetherimide
See Polyimide

Polyethers
The large family of linear heterochain polymers containing the –C–O–C– ether group. The most important members are the thermoplastics polyoxymethylene (acetal); poly(ethylene oxide) and poly(propylene oxide); poly(phenylene oxide); and the polysulphones PSU, PPSU and polyetheretherketone PEEK. Polyether blocks are important in polyurethane chemistry (*see* page 173) and therefore are found in elastomers and thermosets. The ASTM class O elastomers (including the epichlorohydrin rubbers CO and ECO and propylene oxide rubber GPO) are polyethers. Finally, the sulphur analogues, the thioethers, may be included, the most notable example being poly(phenylene sulphide) PPS.

Polyethersulphone
See Polysulphone

Polyethylene
The polyethylene family of commodity and engineering thermoplastics have an unrivalled importance in the polymer materials industry. PE also has a singular position in polymer science since its uniquely simple molecular structure has made it the object of a vast number of fundamental physical and chemical studies. (*See*, for example, page 37 for discussion of PE crystallinity and morphology.) Although PE has been a commercial material for about 50 years, its materials technology continues to develop extremely vigorously and PE consumption still increases.

The following major groups of polyethylene materials are produced today:

Low-density polyethylene LDPE: the descendant of the original polyethylene produced by ICI around 1939, synthesised by the high-pressure gas-phase free-radical process. Density range 915–930 kg/m^3.
High-density polyethylene HDPE: the linear (unbranched) form of PE first produced at low pressures with Ziegler catalysts but now mainly produced by the Phillips process; includes ethylene copolymers with 1-butene or 1-hexene. Density range: homopolymer, 960–970; copolymers 940–950 kg/m^3.
Linear low-density polyethylene LLDPE: a rival to LDPE in which the degree of crystallinity, the density and the number of small side branches are controlled by deliberate copolymerisation with olefins such as 1-butene, 1-hexene or 1-octene; mainly Ziegler processes. Density range 915–940 kg/m^3. Uniform length of comonomer side-branches compared with LDPE gives rise to some differences in properties: thus LDPE has lower melting temperature and modulus than LLDPE of same density. (For example, at a density of 930 kg/m^3, LDPE melts at 112°C, LLDPE at 125°C; LDPE tensile modulus 300 MPa, LLDPE 400 MPa).
Ultra-high molecular weight polyethylene UHMWPE: a form of PE with exceptionally long, unbranched chains (relative molecular mass as high as 6 × 10^6):

produces a polyethylene with remarkable toughness and abrasion resistance but extremely high melt viscosity.

Polyethylenes may be modified after polymerisation by chlorination, chloro-sulphonation and by chemical or radiation crosslinking.

Copolymerisation with other olefins has also generated materials of commercial value: the most important being the copolymers with propylene, particularly the ethylene–propylene–diene terpolymer elastomers, vinyl acetate (EVA), vinyl alcohol and methacrylic acid (ionomer).

US consumption 1984 (millions of tonnes):

LLDPE	1.2
LDPE	2.1
EVA	0.5
HDPE	2.7.

World production total 1986, 10.0.

Doak, K. W. *et al.*, 'Ethylene polymers', in *Encyclopaedia of Polymer Science and Engineering* vol. 6, 2nd edn, pp. 383–564 (Wiley, New York, 1986)

Poly(ethylene oxide)

A water-soluble polyether, known also as poly(ethylene glycol). PEO is a crystalline thermoplastic (T_g $-55°C$). In solution, small quantities of PEO greatly reduce turbulent friction in shear flow. Addition of 0.01 per cent PEO to water reduces by 80 per cent the friction factor of high Reynolds number turbulent flow in pipes. *Drag reduction* by polymers is not well understood but is believed to be linked with the way in which high-frequency turbulent eddies are modified by the presence of flexible polymer chains. The very high local energy in turbulent flow may cause considerable chain stretching and chain breaking. These are general phenomena of dilute polymer solutions, not specific to aqueous PEO.

de Gennes, P. G., 'Towards a scaling theory of drag reduction', *Physica*, **140A** (1986) 9–25.

Poly(ethylene terephthalate)

PETP is linear polyester, first used as a fibre but now significant also as a thermoplastic material. Thermoplastic PETP achieved real prominence in the 1970s when blow-moulding techniques for bottle manufacture were mastered. PETP preforms are biaxially oriented in the blow-moulding process, producing a highly transparent, rigid material with low permeability to water and carbon dioxide. Since 1978, the growth of the market for PETP bottles has been spectacular.

PETP suffered as a moulding resin from the problem of very slow crystallisation (*see* page 48); these problems were ingeniously overcome in the late 1970s by Du Pont who stimulated the nucleation process chemically by blending PETP with a small quantity of ionomer and lowering the glass transition temperature with a plasticiser. Glass-fibre reinforced PETP is now a well-established moulding resin with a world market.

Polyimide

A group of linear aromatic polymers resistant to very high temperatures but rather intractable in processing. Many polyimide chain structures have been investigated (see table 5.7 for a typical structure). Both thermoplastic and thermosetting polyimides have been commercialised, at present in rather small quantities. The polyimides are thermally stable (in the absence of oxygen) to around 500°C.

Polyetherimide PEI is a related polymer in which the imide structure in the chain is diluted with aromatic polyether units, producing greater chain flexibility and somewhat improved processing properties. Thermoplastic PEI (heat deflection temperature 220°C) was commercialised by General Electric in 1982. A closely related polyamideimide with an aromatic amide group in the chain repeat unit has been developed by Amoco.

Mittal, K. L. (Ed.), *Polyimides, Synthesis, Characterisation and Applications*, vols 1 and 2 (Plenum, New York, 1984).

Polyisobutylene

A carbon-chain homopolymer, a close structural relation of butyl rubber, poly(iso-butylene-*co*-isoprene) IIR. The low chain length polyisobutylenes are polymer liquids, the high chain length materials are elastomers. Long chain polyisobutylene crystallises on stretching, the chains adopting the spectacular helical structure shown in figure 2.4(c).

Polyisoprene

Synthetic poly(*cis*-1,4-isoprene) IR, structurally identical to natural rubber, is synthesised with Ziegler catalysts. IR and NR are very similar in properties. IR production capacity is about 1 million tonnes worldwide, approximately 10 per cent of total synthetic rubber.

Senyek, M. L., 'Isoprene polymers', in *Encyclopaedia of Polymer Science and Engineering*, vol. 8, 2nd edn, pp. 487–564 (Wiley, New York, 1987).

Polymethylpentene

A hydrocarbon polymer unusual in being both crystalline and transparent. It is a member of the polyethylene, polypropylene series, produced in only small quantities. Its transparency probably arises from the fact that crystalline and amorphous regions have the same density and hence the same refractive index. The polymer has the lowest density of any commercial thermoplastic (0.83), lower even than PP; it has a high melting point (250°C) but dissolves rather easily in hydrocarbon solvents.

Polynorbornene

A recently introduced hydrocarbon elastomer of very unusual chain structure: the first polymer to incorporate a five-carbon ring in the main chain.

Polynorbornene has a high glass transition temperature, 35°C, and is heavily plasticised with hydrocarbon oils to develop elastomeric properties at ambient temperatures.

Polyolefins

Olefin is the chemist's name for the unsaturated hydrocarbons (such as ethylene, propylene, butylene and so on) which are the monomers from which such important polymers as polyethylene, polypropylene and polyisobutylene are produced, commonly today by stereospecific addition polymerisation using Ziegler–Natta type catalysts.

Polyoxymethylene

A linear heterochain polymer of very simple primary chain structure (*see* page 13). An easy polymerisation was known in the nineteenth century but POM was not produced commercially until about 1959 when a means was found by Du Pont to overcome the troublesome thermal instability of the polymer which made melt-processing unsatisfactory. The solution lay in chemical modification of the ends of the chains. An acetal copolymer with slightly lower processing temperature (introduced by Celanese in 1962) competes with the homopolymer.

POM is interesting to polymer scientists because of its simple primary chain structure which suggests a natural comparison with polyethylene.

	POM homopolymer	*linear PE*
Crystallinity (%)	65	80–90
Density (kg/m^3)	1420	950
Melting temperature (°C)	180	138
T_g(°C)	−90/−10	−80/−90
Relative permittivity	3.6–4.0	2.3
Heat of combustion (kJ/kg)	16.9	46.5
Tensile modulus (GPa)	2.6–3.4	0.4–1.0
Tensile strength (MPa)	65–72	18–33

POM is stiffer, stronger and higher melting than linear PE; its more polar chain structure makes the intermolecular lattice energy somewhat greater. On the other hand, the polar bond makes the polymer less resistant to attack by acids.

Dolce, T. J. and Grates, J. A., 'Acetal resins', in *Encyclopaedia of Polymer Science and Engineering*, vol. 1, 2nd edn, pp. 42–61 (Wiley, New York, 1985).

Polypeptide

A class name for the biopolymers formed by combination of naturally occurring amino acids. The repeat unit in the chain is –CHR–CO–NH–, similar to that of the synthetic polyamides. Whereas the synthetic polymer is formed by polymerisation of a single amino acid (or the copolymerisation of a diacid and a diamine), biological polypeptides are formed by the controlled polymerisation of some 20 naturally occurring aminoacids in precise sequences determined directly by the genetic material of the cell. Natural polypeptides fall broadly into two groups: the fibrous proteins (notably keratin and collagen) which serve as structural materials; and the globular proteins which have a metabolic function, notably as enzymes. The regular structure of peptide links in the main chain of the macromolecule allows a helical structure stabilised by H-bonds to develop in most proteins. In globular proteins, the side-groups R which each aminoacid contributes to the linear chain interact, through H-bonds and disulphide bonds, to produce elaborate twisting and folding of the helical macromolecular chain. This overall chain conformation is of course identical in each molecule of a particular polypeptide since each has the same aminoacid sequence. Consequently, many globular proteins can be crystallised, the unit cell being built from entire globular macromolecules (rather than the chain repeat units as in the crystalline synthetic polymers described in chapter 2).

Poly(phenylene ether) –
See Poly(phenylene oxide)

Poly(phenylene oxide)

An essentially amorphous linear heterochain polymer developed by General Electric in the mid 1960s. PPO has an exceptionally high glass transition temperature of 208°C (creating processing problems) but has the remarkable property of being miscible in all proportions with polystyrene. The PPO–PS hybrids are therefore one-phase blends. They have established themselves as among the most successful of the new engineering thermoplastics. PPO is also modified by copolymerisation with a polyamide.

Poly(phenylene sulphide)

A linear crystalline thermoplastic first produced by Phillips in 1973. It is a sulphur heterochain near-analogue of PPO. In the very simple molecular structure, the six-carbon (aromatic) rings provide chain stiffness and temperature stability. The melting point is around 285°C and the glass transition temperature around 85°C. Temperatures as high as 350°C are necessary for injection moulding. An oddity of PPS is that the raw polymer can be 'cured' by heat-treatment at 150–250°C, a process believed to cause an increase of chain length and perhaps some crosslinking. This thermoplastic therefore has some thermoset character.

Brady, D. G. 'Poly(phenylene sulphide)', *J. Appl. Polymer Sci.: Appl. Polymer Symposia*, **36**, (1981) 231-239.

Hill, H. W. and Brady, D. G. 'Poly(phenylene sulphide)', in M. Grayson (Ed.), *Kirk-Othmer Encyclopaedia of Chemical Technology*, 3rd edn (Wiley, New York, 1984).

Polypropylene

PP has a special place in the history of polymers. The emergence of isotactic PP in 1954 was unexpected yet so logical, given our present understanding of the molecular and structural basis of polymer materials. Furthermore, the scientific triumph of Guilio Natta was translated within three years into a commercial triumph when Montecatini brought PP to the marketplace in 1957. The properties of solid isotactic PP showed very vividly the rewards which might be gained by control of stereoregularity during synthesis. It was exceptional that an outstanding contribution to the science of catalytic polymerisation should also yield a polymer material of great commercial value; and that the laboratory synthesis could be so effectively transferred to the industrial scale. Moreover, the original success has been followed by a sustained development both of catalysts and of PP materials and applications. The original PP homopolymer has now spawned a variety of copolymers and hybrids which together form the PP family (*see* page 156 for a description of these).

Galli, P., 'Polypropylene: a quarter of a century of increasingly successful development', in F. Ciardelli and P. Giusti (Eds), *Structural Order in Polymers*, pp. 63-92 (Pergamon, Oxford, 1981).

Polystyrene and other styrene polymers

Polystyrene is a commodity polymer with wide application but many deficiencies in end-use properties, especially in its susceptibility to brittle fracture. Styrene is a very versatile monomer and is very widely hybridised with other substances to form important copolymers, blends and alloys. High-impact polystyrene HIPS and acrylonitrile-butadiene-styrene ABS were the first such hybrids (*see Acrylonitrile-butadiene-styrene*). In both these materials, the presence of a grafted rubbery disperse phase radically modifies the propagation of fractures and hence improves impact strength. Crazes which develop around the rubber particles under fracture stress absorb energy and raise the fracture toughness. More recently, the interesting combination of PPO-PS has been developed as a high-performance engineering polymer.

Polysulphide elastomers

Poly(ethylene polysulphide) is an interesting rubber with an unusual chain structure containing sulphur-sulphur bonds. It has unique resistance to swelling by organic solvents. However it is a weak material and is mainly used as a sealant and adhesive.

Polysulphone
Linear heterochain polymers containing the $-SO_2-$ group. The first commercial polysulphone was introduced by Union Carbide in 1965 (*see* table 1.5 for structure); polyethersulphone, with a rather simpler chain structure, was commercialised by ICI in 1971:

Typically the polysulphones are tough, glassy polymers with T_g in the region of 200–250°C. They are very stable to thermal oxidation and to creep and are suited to prolonged use under stress at high temperatures. They can be injection moulded and extruded at melt temperatures around 350°C. PES has a UL thermal index of 180°C.

Polytetrafluorethylene
The fully fluorinated analogue of polyethylene, which was unexpectedly discovered by Roy Plunkett of Du Pont in 1938 in the course of his work on refrigeration fluids. PTFE is a polymer of remarkable properties, many attributable to the stability of the C–F bond. It is outstandingly useful as a high-frequency dielectric; has an extraordinarily low coefficient of friction; is resistant to chemical attack by almost all substances; and resists atmospheric degradation. It presents some difficulties of processing (*see* page 165) because of its high melting temperature (around 325°C) and very high melt viscosity, attributable to the high molar mass and relatively inflexible linear chain (*see* section 2.8). PTFE came into commercial production in 1950, although there were military uses from about 1943. Several other fluorocarbon polymers (notably PCTFE and the tetrafluorethylene-hexafluoropropylene copolymer FEP) offer most of the advantages of PTFE combined with better melt processing characteristics.
Sperati, C. A., 'Polytetrafluorethylene: history of its development and some recent advances', in R. B. Seymour and G. S. Kirshenbaum (Eds), *High Performance Polymers: their Origin and Development*, pp. 267–278 (Elsevier Applied Science, London, 1986).

Polyurethanes
A most important and diverse class of chemically complicated polymers including thermoplastic, thermoset and elastomer materials. For a brief survey of PUR elastomer chemistry, *see* table 6.7. PUR materials are very widely used as foams, both rigid and flexible; reaction injection moulding methods were developed principally for PUR materials. 1985 US consumption of polyurethanes was

about 1.0 million tonnes, similar to that of phenolics and just under one half
of that of polypropylene.

Hepburn, C., *Polyurethane Elastomers* (Applied Science, London, 1982).
Oertel, G., *Polyurethane Handbook* (Hanser, Munich, 1985).
Woods, G., *Flexible Polyurethane Foams* (Applied Science, London, 1982).
Woods, G., *The ICI Polyurethanes Book* (Wiley, Chichester, 1987).

Poly(vinyl acetate)

A vinyl polymer produced by emulsion or suspension polymerisation as a latex
and widely used in coatings and adhesives. Like PVC it may be plasticised, often
temporarily in emulsion coating applications.

Poly(vinyl alcohol)

A water-soluble vinyl polymer, produced by hydrolysis from PVAC. Water
solubility arises from strong H-bond interaction between hydroxyl groups on
the polymer chain and water molecules.

Poly(vinyl butyral)

A vinyl polymer produced by post-polymerisation reaction of PVAL with
butyraldehyde. This reaction removes most of the hydroxyl groups from the
carbon chain, rendering the polymer insoluble in water. Poly(vinyl butyral)
has one predominant use: as the bonding layer in vehicle safety glass, for which
purpose it has a uniquely effective combination of stability, clarity and glass-
adhesion.

Poly(vinyl chloride)

One of the cheapest and most versatile commodity polymers, surpassed only by
polyethylene in production volume. Much of the PVC produced finds its way
into the construction industry. Unplasticised PVC has a glass transition temperature of about 85°C. Commercial PVC is generally regarded as largely atactic with
sufficient syndiotactic sequences in its chain structure to allow slight crystallinity
to develop. PVC has a unique capacity to absorb a wide variety of organic
solvents which act as plasticisers and much of PVC technology centres on the
modification of PVC material properties by plasticisation, lowering T_g and
making the material more flexible. PVC decomposes readily above 200°C to
yield large amounts of hydrogen chloride gas. Commercial formulations are
stabilised with lead, cadmium, tin or antimony compounds to reduce thermal
decomposition during processing. PVC latices and organosols with controlled
particle size are produced by suspension and emulsion polymerisation and are
much used in the coatings industry. PVC copolymers with small amounts of vinyl
acetate or vinylidene chloride have improved processing properties. PVC may
also be post-chlorinated to produce a commercial thermoplastic more soluble
in organic solvents than PVC itself and with a lower softening point.

Burgess, R. H. (Ed.), *Manufacturing and Processing of PVC* (Applied Science, London, 1982).

Gottesman, R. T. and Goodman, D. 'Poly(vinyl chloride)' in R. W. Tess and D. W. Poehlein (Eds), *Applied Polymer Science*, 2nd edn (American Chemical Society, Washington DC, 1985).

Titow, W. (Ed.) *PVC Technology*, 4th edn (Elsevier Applied Science, London, 1984).

Poly(vinyl fluoride)

A crystalline polymer with exceptional resistance to weathering and chemical attack, used mainly as a coating, especially to control metallic corrosion in aggressive environments. It is the fluorine analogue of PVC, with structure $-(CH_2-CHF)-$.

Poly(vinylidene chloride)

A vinyl polymer widely used in food-packaging because of its extremely low permeability to water vapour (*see Vinyl polymers*).

Poly(vinylidene fluoride)

A crystalline polymer with remarkable piezoelectric properties. On stretching, the PVDF crystal structure changes from the nonpolar alpha to the polar beta form. The electronegativity differences of H and F ensure that the unit cell as a whole has a net dipole. The beta crystallites are oriented by 'poling' the polymer film in a strong electric field (typically 1000 kV/cm) for some minutes at 80°C. The frozen-in polarisation of the material is altered by mechanical deformation, thus providing the piezoelectric property which can be exploited in electromechanical transducers (see page 109). The piezoelectric coefficient $d(= (\partial \epsilon / \partial E)_\sigma$, where ϵ is the strain, E the electric field and σ the stress) of PVDF is about 25 pC/N, much higher than other polymers and comparable with piezoceramics such as barium titanate (190 pC/N).

There are other less exotic uses of PVDF which mainly exploit its combination of inertness, resistance to chemical attack, good fire performance and toughness: for example, in architectural coatings.

Lovinger, A. J., 'Poly(vinylidene fluoride)', in D. C. Bassett (Ed.), *Developments in Crystalline Polymers*, vol. 1 (Applied Science, London, 1982).

Protein
See Polypeptide

Silicone polymers

A diverse group of polymers known primarily for their chemical inertness and relatively good temperature stability. The main structural feature of the silicones is the complete absence of carbon from the chain backbone which instead is built from linear $-Si-O-Si-O-$ sequences. Carbon-containing groups are usually present as side groups attached to the Si atoms. Many silicones are network structures, crosslinked by –Si–O–Si– units. The thermal stability of silicones is attributed to the strength of the Si–O bond. Silicone polymers find numerous relatively specialised uses (they are expensive): in coatings, as surfactants and especially as special-purpose elastomers for aggressive, high-temperature and low-temperature environments. Silicone elastomers may be heat-cured with peroxides but especially important are the *room-temperature vulcanised* (*RTV*) silicone rubbers. These are one or two-pack systems which cure in the cold and are important as adhesives and sealants, especially in the construction industry. They have good weathering performance and remain flexible at low temperatures. The partially fluorinated fluorosilicone rubbers combine the chemical inertness of the silicones with even better flexibility at low-temperature.

Cush, R. J. and Winnan, H. W., 'Silicone rubbers', in A. Whelan and K. S. Lee (Eds), *Developments in Rubber Technology*, vol. 2, pp. 203–231 (Applied Science, London, 1981).

Warrick, E. L., Pierce, O. R., Polmanteer, K. E. and Saam, J. C., 'Silicone elastomer developments', *Rubber Chem. Technol.*, **52** (1979) 437–525.

Styrene polymers
See *Acrylonitrile-butadiene-styrene, Polystyrene and other styrene polymers and Styrene–butadiene rubber*

Styrene–butadiene rubber
The styrene–butadiene random copolymer elastomer is the most important of the synthetic rubbers (world production 1986, around 5 million tonnes, 60 per cent of total synthetic rubbers), exceeding natural rubber in production volume. SBR competes with natural rubber in many applications, particularly in vehicle tyres. Having a random chain structure, it does not crystallise at high extensions; it requires reinforcing fillers to develop good mechanical properties. Styrene–butadiene block copolymers have been developed as important thermoplastic elastomers (*see Thermoplastic elastomers*).

Thermoplastic elastomers
A large class of polymers which, by Hoffmann's excellent definition, "in the ideal case, combine the service properties of elastomers with the processing properties of thermoplastics." These are broadly those rubbers with the ASTM class Y designation ("polymers having a block, graft, segmented or other structure that permits use of the rubber in the unvulcanized state at ordinary temperatures; for example YSBR. . ." [ASTM D1418]. These materials develop elastomeric properties at normal temperatures because of their blocky two-phase structure (*see* pages 173–175 for discussion). There are numerous commercial thermoplastic elastomers, which may be classified as follows (after Hoffmann):

Components	Soft block	Hard block
Styrene types		
SBS, SIS	butadiene or isoprene	styrene
SEBS	ethylene/butadiene	styrene
Elastomeric alloys		
EPDM/PP	crosslinked EPDM	PP
NR/PP	crosslinked natural rubber	PP
EVA/PVDC	EVA	PVDC
NBR/PP	crosslinked NBR	PP
Polyurethanes	polyester	polyurethane
	polyether	polyurethane
Polyetheresters	alklyene glycol	alkylene terephthalate
Polyetheramides	etherdiols	amides

Hoffmann, W., 'Thermoplastic elastomers: classes of material and an attempt at classification', *Kunststoffe German Plastics*, 77(8) (1987) 14–20.

Holden, G., 'Thermoplastic elastomers', in *Encyclopaedia of Polymer Science and Engineering*, vol. 5, 2nd edn, pp. 416–430 (Wiley, New York, 1986).

Legge, N. R., Holden, G. and Schroeder, H. E., *Thermoplastic Elastomers: a Comprehensive Review* (Hanser, Munich, 1987).

Vinyl polymers

The family of polymers comprising poly(vinyl chloride), its copolymers with vinyl acetate, vinylidene chloride, ethylene and propylene; poly(vinyl acetate); poly(vinylidene chloride); poly(vinyl alcohol); and poly(vinyl butyral).

Index of Subjects

230

Index of Authors